Lecture Notes in Control and Information Sciences

Edited by A. V. Balakrishnan and M. Thoma

For information about Vols. 1–21 please contact your bookseller or Springer-Verlag.

Lecture Notes in Control and Information Sciences

Edited by M. Thoma and A. Wyner

86

Time Series and Linear Systems

Edited by S. Bittanti

Springer-Verlag Berlin Heidelberg GmbH

Editor
Sergio Bittanti
Dipartimento di Elettronica
Politecnico di Milano
Piazzo Leonardo da Vinci 32
20133 Milano (Italy)

Library of Congress Cataloging in Publication Data
Time series and linear systems.
(Lecture notes in control and information sciences; 86)
Includes bibliographies.
1. Time-series analysis. 2. Linear systems.
I. Bittanti, Sergio. II. Series.
QA280.T558 1986 519.5'5 86-20244

ISBN 978-3-540-16903-1 ISBN 978-3-540-47155-4 (eBook)
DOI 10.1007/978-3-540-47155-4

Originally published by Springer-Verlag Berlin Heidelberg New York in 1986

2161/3020-543210

PREFACE

Over the five past years, a stream of research at the Politecnico di Milano (Italy) has been in the methodology of modelling and identification of time series by means of linear systems. Several specialists of different backgrounds, including System and Control Theory, Statistics, Econometrics, Numerical Analysis, visited the Politecnico contributing with their talks to setting up a workshop on the subject. The train of ideas underlying this activity was to develop a system-theoretic point of view for the art of modelling.

This book is a partial report of such an activity. The various chapters are extended introductory papers overviewing important advanced topics in the field. They also constitute useful introductions to research directions of current interest.

The book is organized as follows. The first chapter is an introduction to the use of stochastic models in time series analysis. The problem of modelling is interpreted here as the problem of finding the linear model which is the best approximant for the data at hand. Among other things the use of criteria such as AIC or BIC is critically discussed. Moreover, the problem of determining a suitable rational transfer function approximation is studied as the problem of approximating the infinite Hankel matrix of the impulse response coefficients with a Hankel matrix of finite rank. Linear systems where all observed variables are subject to errors are considered in the second chapter. The motivation is that prejudicial causality assumtions can then be avoided. A new class of dynamic models for time series is proposed in the third chapter. These models are based on the classical Factor Analysis approach, and are strictly related to the systems introduced in Chapter 2. The fourth chapter is devoted to the so called Minimum Description Length approach. A model is then judged by the number of binary digits with which it permits to encode the observed data. This leads to the notion of stochastic complexity of the data, as the shortest number of binary digits with which it permits to encode the observed data. Chapter 5 deals with systems with periodically time-varying coefficients, which can be used to describe seasonal time series. The attention focuses here on the basic structural properties of these systems, i.e. reachability, stabilizability and so on. The role played by these properties in the analysis of stochastic periodic systems is touched upon. Some numerical problems in linear system theory are considered in the sixth chapter. An extensive overview of the LU, QR, Shur and Singular Value Decomposition algorithms is provided. Then, the problem of computing the reachability subspace of a time-invariant system is studied. The last chapter is devoted to the discussion of some recent trends in Econometrics.

The volume can be used either as a textbook for monographic courses on the subject or as a reference book providing researchers with the main trends and perspectives in the field.

The editor expresses his sincere acknowledgment to the fellow authors for their most valuable contributions, as well as their care and patience in the preparation of the manuscripts.

The support of the Centro di Teoria dei Sistemi of the National Research Council (C.N.R.) and that of the Ministry of Education (M.P.I.) is gratefully acknowldged.

Sergio Bittanti

ABSTRACTS

Chapter 1 TIME SERIES AND STOCHASTIC MODELS

by E.J. Hannan

The basic concept of this paper is a linear system wherein an output $y(t)$, of q components, is related to an input, $u(t)$, of p components via a relation

$$y(t) = \sum_0^\infty W_i \, e(t-i) + \sum_1^\infty L_i \, u(t-i)$$

wherein the $e(t)$ are the linear prediction errors for $y(t) - \Sigma L_i \, u(t-i)$. The methods of the paper are substantially valid when the system is truly linear in the sense that linear prediction is optimal, but may prove useful over a much wider range.

To bring the problem back to reasonable proportions the statistical methods are based on the approximation of the true structure by one wherein the matrix functions

$$W(z) = \Sigma W_i \, z^{-i}, \; L(z) = \Sigma L_i \, z^{-i}$$

are approximated by matrices of rational functions. A brief discussion is given of some basic theory relating to such an approximation process. It is necessary, in the approximation to choose the "order" of the approximant, i.e. effectively the maximum lags in the ARMAX model,

$$\sum_0^h A_i y(t-i) = \sum_1^h B_i u(t-i) + \sum_0^h C_i \, e(t-i) \, ,$$

to which the rational transfer function corresponds.
Various algorithms are described that are basic in time series analysis and are then used to effect a solution to the problem of finding a suitable approximant. The main algorithm described does this by a Gauss-Newton iteration in which the order is redetermined at each iteration by a calculation recursive in the order.
Finally, on-line implementations of the algorithm are presented for the case where $y(t)$ is scalar.

Chapter 2 LINEAR ERRORS-IN-VARIABLES SYSTEMS
by M. Deistler

Linear errors-in-variables(EV) systems, i.e. linear systems where all observed variables are subject to errors are considered. The statistical analysis of such systems turns out to be significantly more complicated compared to conventional errors in equations (e.g. ARMAX) systems. A good part of these complications arises from the fact that the transfer function of the system in the EV case, in general, is not uniquely determined from the second moments of the observations.
The paper is organized as follows: In section 2 some well known results concerning the static case are restated. In sections 3 - 5 the information about the transfer function contained in the (ensemble) second moments of the observations is analysed: In section 3 the set of all transfer functions

corresponding to given second moments of the observations is described. Section 4 deals with the same problem when the system is a priori known to be causal and with the problem whether causality can be detected from the second moments of the observations. In section 5 conditions for identifiability are derived. Section 6 deals with conditions for identifiability using information coming from moments of order greater than two.

Chapter 3 A NEW CLASS OF DYNAMIC MODELS FOR STATIONARY TIME SERIES

by G. Picci and S. Pinzoni

A new class of dynamic models for stationary time series is presented. They are a natural generalization of the well-known linear *Factor Analysis* Models widely used in Statistics and Psychometrics. It is shown that the Factor Analysis Models of time series considered in this note reduce to (and to some extent clarify the structure of) *Dynamic Errors-In-Variable Models* discussed in the recent literature. They provide simple mathematical schemes for the identification of multivariate time series which avoid the unjustified introduction of a priori *causality* assumptions as for example subsumed by conventional ARMAX models.

Chapter 4 PREDICTIVE AND NONPREDICTIVE MINIMUM DESCRIPTION LENGTH PRINCIPLES

by J. Rissanen

This chapter presents in a tutorial manner the basic ideas behind the recently developed estimation principle, called Minimum Description Length principles. Briefly, a statistical model is judged by the number of binary digits with which it permits one to encode the observed data. The shortest code length available for models in a class is defined to be the stochastic complexity of the data. Depending on how the coding is done two kinds of stochastic complexities can be defined, the predictive and the nonpredictive ones, which for large samples tend to the same value. The stochastic complexity also sets a tight lower bound for the errors with which the data can be predicted. The model associated with the complexity involves estimates both of the number of the parameters and their values, which may be taken to incorporate all the statistical information that can be extracted from the data with the considered models. Hence, we may say that the fundamental problems in modeling are to calculate the stochastic complexity and the associated optimal model.
As applications we describe the calculation of the stochastic complexity of the data relative to the gaussian ARMA class of models, both in the single and the multiple input/output case.We illustrate with simulations the consistency of the associated estimates of the number of the parameters and the structures. We also describe how the prior knowledge about the parameters, as represented by their estimated values, can be taken advantage of. The feasibility of the scheme is demonstrated by simulations.

Chapter 5 DETERMINISTIC AND STOCHASTIC LINEAR PERIODIC SYSTEMS

by S. Bittanti

The main results concerning the structural properties of linear periodic systems are reviewed. Both continuous-time and discrete-time systems are dealt with. By a comparison with time-invariant systems, five structural properties are discussed. Three of them are basic properties concerning the reachability and controllability subspaces. The fourth one concerns the length of the time interval required to perform the reachability and controllability transition. The modal (spectral) characterizations are presented as fifth property. The extended structural properties (i.e. stabilizability and detectability) are also dealt with. Finally, periodic stochastic systems are considered. The existence of a cyclostationary solution is investigated by analizing the appropriate periodic Lyapunov equation.

Chapter 6 NUMERICAL PROBLEMS IN LINEAR SYSTEM THEORY

by D. Boley and S. Bittanti

We discuss some numerical aspects in linear system theory.We start by showing the numerical algorithm to solve systems of linear equations and non-degenerate least squares problems.We then move on to an introduction to more sophisticated matrix decompositions, used to solve more sophisticated problems,and introduce the cincept of *backward error analysis* (Wilkinson, 1965). Among the decompositions we introduce

name	form	used to obtain
LU (Gaussian Elimination)	A=LU	. solution of linear Equations . determinant
QR (orthogonal triangularization)	A=QR	. soln. to least Squares problem (linear non degenerate) . soln. to linear Equations without need to pivot
Schur	$A=QRQ'$	. Eigenvalues/vectors
Singular Value Decomposition (S.V.D.)	$A=P\Sigma Q'$	. Singular Values . rank . distance to singularity . 2-norm of matrix . 2-norm condition number

where P,Q denote orthogonal matrices

U,R " upper triangular matrices

L " lower triangular matrices

Σ is non-negative diagonal

In the last section we discuss some numerical aspects in linear system theory. The attention is focused on the problem of computing the controllable subspace of a time-invariant linear system. It is shown how some classical methods lead to numerical problems and give some recent results giving bounds on the errors in terms of results from these classical methods.

<u>Chapter 7</u> SOME RECENT DEVELOPMENTS IN ECONOMETRICS
by M. McAleer and M. Deistler

In this paper we discuss some of the main recent developments in econometrics: methods for specification search, in particular, diagnostic checking and specification testing; macroeconomic modelling and forecasting; and some models associated with empirical microeconomics.

AUTHORS

Sergio Bittanti
Dipartimento di Elettronica
Politecnico di Milano
Piazza Leonardo da Vinci, 32
20133 MILANO
ITALY

Daniel Boley
Department of Computer Science
University of Minnesota
136 Lind Hall
207 Church Street S.E.
MINNEAPOLIS, Minnesota 55455
U.S.A.

Manfred Deistler
Institut für Ökonometrie und
Operations Research
Technische Universität Wien
Argentinierstrasse 8/119
A-1040 WIEN
AUSTRIA

Edward G. Hannan
Department of Statistics
Mathematical Sciences Bldg.
The Australian National University
GPO Box 4 CANBERRA, ACT 2601
AUSTRALIA

Michael J. McAleer
Department of Statistics, The Faculties
The Australian National University
GPO Box 4 CANBERRA, ACT 2601
AUSTRALIA

Giorgio Picci
Istituto di Elettrotecnica ed Elettronica
Università di Padova
Via Gradenigo 6/A
35131 PADOVA
ITALY

Stefano Pinzoni
LADSEB-CNR
Corso Stati Uniti 4
35020 PADOVA
ITALY

Jorma Rissanen
IBM-RES
650 Harry Road
SAN JOSE, CA 95193
U.S.A.

TABLE OF CONTENTS

Chapter 1

Time Series and Stochastic Models

E.J. Hannan

1. Introduction

This chapter will be concerned with procedures for analysing data, $y(t)$, $t = 1,2,\ldots,T$, where $y(t)$ is a vector of q components that can be thought of as the output of some system to which the input is $u(t)$, an observed vector of p components. The situation held in mind is one where no very precise information is available about the system and the description will be on the basis of models of such generality that experience suggests will suffice for a good explanation. This will be further discussed below. These models will always be stochastic.

Let us begin by considering $y(t)$, alone, as generated by a stationary stochastic process with finite mean square, so that

$$E\{y_j(t)^2\} < \infty, \quad j = 1,2,\ldots,q,$$

where $y_j(t)$ is the j'th component of $y(t)$. It is costless to assume that $y(t)$ is ergodic since only one history or realization of the process is ever seen and reasonable to require that it be purely non-deterministic, so that there is no influence on $y(t)$ from the indefinitely far past, or rather if there is such an influence it can only be through such effects as the mean or of periodic components such as diurnal or seasonal movements. Such effects could first be removed by regression so that, for example, all calculations will be with the mean corrected quantities $y(t) - \bar{y}$,

$$\bar{y} = \frac{1}{T}\sum_{1}^{T} y(t).$$

In relation to calculations it will be assumed that this has already been done so that $y(t)$ is the residual from such an adjustment. This makes notation simpler.

Any such stationary, non-deterministic process can be analysed, at least in part, through its spectrum, $f(\omega)$, a $q \times q$ matrix valued function satisfying $f(\omega) = f(\omega)^* = f(-\omega)'$ and

$$\Gamma(t) \stackrel{d}{=} E\{y(s)y(s+t)'\} = \int_{-\pi}^{\pi} e^{it\omega} f(\omega)\, d\omega.$$

We shall not discuss Fourier methods in any detail because the main methods of this paper are different. Here "finite parameter" models are emphasised in contrast to Fourier methods that are non-parametric and in which the generality is reduced to manageable proportions, essentially, by smoothness requirements for $f(\omega)$. These finite parameter models have been especially emphasised in econometrics and systems engineering and are called ARMAX, an acronym for autoregressive moving-average with exogenous components. (Here exogenous means input.) For $y(t)$ stationary and non-deterministic,

$$y(t) = \sum_0^{\infty} W_i\, e(t-i), \quad W_0 = I_q, \quad E\{e(s)e(t)'\} = \delta_{st}\Omega.$$

Here the $e(t)$ are the linear innovations i.e. $e(t) = y(t) - \hat{y}(t|t-1)$ where $\hat{y}(t|t-1)$ is the best linear predictor of $y(t)$ from $y(t-1), y(t-2)\cdots$. There is an extensive theory due to Kolmogoroff, Wiener and others concerning the construction of $\hat{y}(t|t-1)$ from knowledge of $f(\omega)$ but this will not be important, algorithmically, here. Put

$$W(z) = \sum_0^{\infty} W_i z^{-i} \tag{1.1}$$

Then $\det W(z) \neq 0$, $|z| > 1$ and $W(z)$ is analytic for $|z| \geq 1$, since $\Sigma \| W_i \|^2 < \infty$. However we always assume $\det W(z) \neq 0$, $|z| \geq 1$, since zeros on $|z| = 1$ cause considerable problems. There is a decomposition

$$f(\omega) = \frac{1}{2\pi} W(e^{-i\omega})\, \Omega W(e^{-i\omega})^*, \tag{1.2}$$

which is unique since there is no other such decomposition with $W(z)$ having the properties stated above. To take account of $u(t)$ we generalise (1.1) to

$$y(t) = \sum_0^{\infty} W_i e(t-i) + \sum_1^{\infty} L_i u(t-i), \tag{1.3}$$

and put $L(z) = \Sigma L_i z^{-i}$ The essential restriction here is that the relation is causal, so that there is no influence on $y(t)$ from $u(s)$, $s \geq t$. However (1.1), (1.2), (1.3) are too general to serve as a basis for a worthwhile statistical analysis. To introduce a further specialisation consider the infinite (Hankel) matrix

$$\mathcal{H} = \begin{bmatrix} W_1 & L_1 & W_2 & L_2 & W_3 & L_3 & \cdots \\ W_2 & L_2 & W_3 & L_3 & W_4 & L_4 & \cdots \\ W_3 & L_3 & W_4 & L_4 & W_5 & L_5 & \cdots \\ \vdots & \vdots & \vdots & \vdots & \vdots & \vdots & \vdots \end{bmatrix}$$

Here $[W_j L_j]$ will be called a "block", of q rows and $p+q$ columns. The importance of $\mathcal{H}$ can be seen from the, almost obvious, fact that the best j step ahead predictor is, ignoring prediction of $u(t)$,

$$\hat{y}(t+j|t) = \sum_0^\infty W_{i+j}\, e_{t-i} + \sum_0^\infty L_{i+j}\, u_{t-i}, \quad j \geq 1$$

so that $\mathcal{H}$ has, as the j'th row of blocks, the coefficient blocks in that prediction. The importance of $\mathcal{H}$ will be made evident in other ways in section 4.

Let H_0 be a set of n rows of $\mathcal{H}$ that span all of the rows of $\mathcal{H}$ so that any row can be linearly represented in terms of them. Of course n would be infinite in general. The integer n, the rank of $\mathcal{H}$, is called the order or the McMillan degree of $\mathcal{H}$, or equivalently of $[W(z), L(z)]$. Call H_1 the first block of q rows of $\mathcal{H}$ and put
$H_0 = [K\ L\ H_2]$ where K, L comprise, respectively the first q and the next p columns of H_0. Put

$$\mu(t) = [e(t)'u(t)'e(t-1)'u(t-1)'\ldots]', \qquad x(t) = H_0\mu(t-1). \qquad (1.4)$$

Then from (1.3), (1.4)

$$y(t) = H_1\mu(t-1) + e(t), \qquad H_0\mu(t) = Ke(t) + Lu(t) + H_2\mu(t-1).$$

Since H_0, H_2 are composed of (full) rows of $\mathcal{H}$ then $H_1 = HH_0$, $H_2 = FH_0$ for suitable F, H and

$$y(t) = Hx(t) + e(t), \qquad x(t+1) = Fx(t) + Lu(t) + Ke(t). \qquad (1.5)$$

This is the state space representation in prediction error form. Its lack of uniqueness, given that F is minimal, i.e. of dimension the rank of $\mathcal{H}$, is entirely due to the lack of uniqueness in a choice of H_0. That can be made unique by choosing the rows of H_0 as the first linearly independent set found as you go down the rows of $\mathcal{H}$. We will return to this later.

The methods used herein are dependent on acting as if n is finite. Then, and only then, $W(z)$ and $L(z)$ are matrices of rational functions of z and can thus be written in the form

$$[\, W(z)\ L(z)] = A(z^{-1})^{-1}[C(z^{-1})\ B(z^{-1})] \tag{1.6}$$

where $A(z)$, $B(z)$, $C(z)$ are matrices of polynomials. Of course (1.6) is far from unique but we shall later describe how the unique prescription of H_0 just described leads to a unique "matrix fraction description", (1.6). We shall use z^{-1} also to indicate the shift operator i.e. $z^{-1}y(t) = y(t-1)$. Corresponding to (1.6) we have the ARMAX representation

$$A(z^{-1})y(t) = B(z^{-1})u(t) + C(z^{-1})e(t). \tag{1.7}$$

This is important partly because it expresses $y(t)$ in terms of $y(t-1), y(t-2),.. u(t-1), u(t-2),..e(t), e(t-1), e(t-2),..$ and can serve as a basis for an iterative estimation procedure where the coefficient matrices are estimated by regression, the $e(t)$, which are unobserved, being replaced by estimates from a previous stage in the iteration. This will be dealt with in section 5. When no input (or exogenous) variable is observed we speak of the ARMA case.

Notes on References. There are many references for the basic spectral theory, for example Hannan (1970). For the structure theory of systems see Kailath (1980), Casti (1977).

2. Some Basic Algorithms

There are three basic algorithms of time series analysis.

(i) The first algorithm is the discrete Fourier transform

$$y(t) \to d(\omega) = T^{-\frac{1}{2}} \sum_{1}^{T} y(t)e^{it\omega},$$

which is cheaply computable at frequencies $2\pi j/T'$, $j=0,1,\dots,[\frac{1}{2}T']$ for T' highly composite. Under smoothness conditions on $f(\omega)$

$$E\{d(2\pi j/T)d(2\pi k/T)^{*}\} \doteq \delta_{jk} 2\pi f(2\pi j/T)$$

and, indeed the error is uniformly $O(T^{-1})$ if n is finite. This algorithm will not be so important to us.

(ii) The second algorithm is the Levinson-Whittle recursion. This is designed, in a sense, to produce approximations to $e(t)$ in (1.1) by considering

$$e(t) = \Phi(z)y(t) \qquad \Phi(z) = W(z)^{-1}.$$

The procedure recursively calculates polynomial approximations $\hat{\Phi}_n$ of degree n to Φ. We have used n because a system for which $\Phi(z)$ is a polynomial of degree n will, in fact, be of McMillan degree n. The recursive calculation uses the data through the natural estimates of $\Gamma(t)$ of the form

$$\hat{\Gamma}(t) = \frac{1}{T}\sum_{s=1}^{T-t} y(s)y(s+t)', \qquad t \geq 0.$$

For $s < 0$ put $\hat{\Gamma}(t) = \hat{\Gamma}(-t)'$. However it will be convenient to put this Levinson-Whittle recursion in a more general setting because of its many uses later. Thus let $v(t)$ be a vector of s components and put

$$\hat{\Gamma}_v(t) = \frac{1}{T}\sum_{s=1}^{T-s} v(s)v(s+t)' = \hat{\Gamma}_v(-t)', \qquad t \geq 0.$$

The recursion calculates matrices $F_{n,j}$, $\tilde{F}_{n,j}$, S_n, $\tilde{S}_n$. If $v(t) = y(t)$ then $F_{n,j}$ is $\hat{\Phi}_{n,j}$, the coefficient of z^{-j} in $\hat{\Phi}_n(z)$, and correspondingly we have an estimate of $e(t)$

$$\hat{e}_n(t) = \sum_0^n \hat{\Phi}_{n,j}y(t-j), \qquad \hat{\Omega}_n = S_n = \frac{1}{T}\sum_1^{T+n} \hat{e}_n(t)\hat{e}_n(t)'. \quad (2.1)$$

The $\tilde{F}_{n,j}$ would, in this case where $v(t) = y(t)$, correspond to "backwards" residuals, $\hat{r}_n(t)$, corresponding to the time reversed process (as distinct from the forwards residuals $\hat{e}_n(t)$). Thus, putting $\tilde{F}_{n,j} = \tilde{\Phi}_{n,j}$ for $y(t) = v(t)$,

$$\hat{r}_n(t) = \sum_0^n \tilde{\Phi}_{n,j}\, y(t-n+j), \qquad \tilde{\Omega}_n = \tilde{S}_n = \frac{1}{T}\sum_1^{T+n} \hat{r}_n(t)\hat{r}_n(t)'. \quad (2.2)$$

We now give the recursive algorithm in terms of $v(t)$.

$$F_{n,j} = F_{n-1,j} + F_{n,n}\tilde{F}_{n-1,n-j}, \tilde{F}_{n,j} = \tilde{F}_{n-1,j} + \tilde{F}_{n,n}F_{n-1,n-j},$$

$$F_{n,0} = \tilde{F}_{n,0} = I_s.$$

$$F_{n,n} = -\Delta_{n-1}\tilde{S}_{n-1}^{-1}, \; \tilde{F}_{n,n} = -\Delta'_{n-1}S_{n-1}^{-1}, \; \Delta_n = \sum_0^n F_{n,j}\hat{\Gamma}_v(j-n-1).$$

$$S_n = (I_s - F_{n,n}\tilde{F}_{n,n})S_{n-1}, \; \tilde{S}_n = (I_s - \tilde{F}_{n,n}F_{n,n})\tilde{S}_{n-1},$$

$$S_0 = \tilde{S}_0 = \hat{\Gamma}_v(0).$$

In case $s = 1$ we have $S_n = \tilde{S}_n$, $F_{n,j} = \tilde{F}_{n,j}$, $j=1,\ldots,n$, so that the algorithm is simplified.

These procedures have severe disadvantages when T is small or, better, when n/T is not small. This is because they are implicitly founded on the Toeplitz assumption that $v(t) = 0$, $-n < t \le 0$ or $T < t \le T+n$. (This is so called because the system of equations for the $F_{n,j}$ for given n has a block Toeplitz matrix, i.e. one with the same elements down any diagonal.) There have been many modifications, often based on calculating, for example, $\hat{e}_n(t)$, $\hat{r}_n(t)$ (see (2.1), (2.2)) recursively by

$$\hat{e}_n(t) = \hat{e}_{n-1}(t) + \hat{\Phi}_{n,n}\hat{r}_{n-1}(t-1), \quad \hat{r}_n(t) = \hat{r}_{n-1}(t-1) + \tilde{\Phi}_{n,n}\hat{e}_{n-1}(t)$$

$$\hat{r}_n(0) \equiv 0, \quad \hat{e}_0(t) = \hat{r}_0(t) = y(t), \quad 1 \le t \le T.$$

Then also

$$\Delta_n = \frac{1}{T}\sum_{1}^{T+n} \hat{e}_n(t)\hat{r}_n(t-1). \tag{2.3}$$

It is the terms in (2.1), (2.2), (2.3) for $T < t \le T+n$ that seem to cause most of the trouble, though those for $1 \le t < n$ also involve, in a substantial way, the Toeplitz assumption. In case $q = 1$ it has been suggested that $\hat{\phi}_{n,n}$ be replaced by

$$\sum_{n}^{T} \hat{e}_{n-1}(t)\ \hat{r}_{n-1}(t-1)/\{\tfrac{1}{2}\sum_{n}^{T} \hat{e}_{n-1}(t)^2 + \tfrac{1}{2}\sum_{n}^{t} \hat{r}_{n-1}(t-1)^2\} \tag{2.4}$$

but one might equally use the coefficient of correlation between $\hat{e}_{n-1}(t)$, $\hat{r}_{n-1}(t-1)$, $n \le t \le T$. A virtue in (2.4) is that the resulting number lies in $(-1,1)$, as also does the correlation coefficient. This is also true of $\hat{\Phi}_{n,n}$ and is completely equivalent to the fact that $\hat{\Phi}_n(z) \ne 0$, $|z| \le 1$, a desirable property. The use of (2.3) involves additional calculations. These lattice or ladder methods (so called because of the flow diagrams describing them) are important in real time calculations. For the purposes of this account we shall continue to write in terms of the Levinson-Whittle recursion but wherever that is used it could be replaced by a lattice formula.

In the scalar case the $\hat{\phi}_{n,n}$ (using a lower case symbol for $q = 1$) are called partial autocorrelations by statisticians and reflection coefficients by systems engineers.

To see why the algorithm has been presented for general $v(t)$ consider computing an estimate of $e(t)$ when inputs are observed.

Put, then,

$$\hat{e}_n(t) = \sum_0^n \hat{\Phi}_{n,j}\, y(t-j) - \sum_1^n \hat{\Psi}_{n,j}\, u(t-j).$$

Here $\Sigma\hat{\Psi}_{n,j}\, z^{-j}$ is an approximation to $W(z)^{-1}L(z) = C(z^{-1})^{-1}B(z^{-1})$. using (1.6). To obtain $\hat{\Phi}$, $\hat{\Psi}$ take $v(t)' = (y(t)', u(t)')$, $s = p + q$, and $[\hat{\Phi}_{n,j}, \hat{\Psi}_{n,j}]$ as the first block of q rows in $F_{n,j}$. Then also $\hat{\Omega}_n$, the covariance matrix of the $\hat{e}_n(t)$, is the top left hand $q \times q$ matrix of S_n. This type of procedure will repeatedly be used below.

(iii) The third major algorithm is the Kalman filter, which computes a <u>finite past</u> equivalent to $e(t)$, on the basis of (1.5), for n finite. The algorithm is

$$\hat{x}(t+1) = F\hat{x}(t) + Lu(t) + K(t)\varepsilon(t), \quad y(t) = H\hat{x}(t) + \varepsilon(t)$$

$$K(t) = \{FP(t)H' + K\Omega\}\{HP(t)H' + \Omega\}^{-1}$$

$$P(t+1) = FP(t)F' + K\Omega K' - K(t)\{HP(t)H' + \Omega\}^{-1}K(t)'$$

$$P(1) = FP(1)F' + K\Omega K', \quad \hat{x}(1) = 0.$$

It may be wise to symmetrise $P(t)$, replacing it by $\frac{1}{2}\{P(t)+P(t)'\}$ to reduce the effects of rounding errors.

There is an enormous literature surrounding this algorithm. For our purposes its importance lies in the fact that it allows the Gaussian likelihood to be calculated, or better $(-2T^{-1})$ by that likelihood, which we call $L(\theta)$ and still speak of as the likelihood. This is, apart from a constant,

$$\frac{1}{T}\sum_1^T \log\det\{HP(t)H'+\Omega\} + \frac{1}{T}\sum_1^T \varepsilon(t)'\{HP(t)H'+\Omega\}^{-1}\varepsilon(t). \tag{2.5}$$

Here θ stands for the parameters involved, i.e. those in F, H, K, Ω. Those in F, H, K we shall call system parameters and shall indicate by τ. The remainder are the variances and covariances in Ω. In (2.5) the Gaussian likelihood has been written down treating $u(t)$ as a fixed sequence of vectors. We emphasise that few, if any, of the methods of this chapter depend greatly on the assumption that the $e(t)$ are Gaussian. The likelihood, (2.5), is used to obtain an estimation method rather than because it is the true likelihood.

<u>Notes on References</u>. The fast Fourier transform was introduced to

latter day science in Cooley and Tukey (1965). The vector form of the Levinson-Whittle algorithm was given in Whittle (1963). Lattice forms are surveyed in Friedlander (1982). A great amount of detail about the Kalman filter is found in Anderson and Moore (1979).

3. Approximation Criteria

The problem to be considered in the remainder of this chapter is that of approximating the true system by one of finite McMillan degree. This degree, n, has to be determined. Once this is recognised it must also be recognised that it is not possible to proceed purely through the minimisation of (2.5) since that can always be further reduced by taking n large. The alternative procedures here considered choose n by minimising some form of

$$\log \det \hat{\Omega}_n + d(n)C_T/T, \qquad n = 0,1,\ldots,N. \tag{3.1}$$

Here $\hat{\Omega}_n$ is the maximum likelihood estimate of Ω, given n, and the first term in (3.1) is, except for a constant, essentially the minimal value of (2.5), for n given. (Some approximation is involved in that statement.) The constant $d(n)$ is the dimension of τ, which is $n(2q + p)$. The second term in (3.1) is a penalty term which increases as n increases, whereas the first decreases. Two commonly used C_T sequences are $C_T \equiv 2$, in which case (3.1) will be called AIC(n), and $C_T = \log T$, in which case (3.1) will be called BIC(n). An upper bound, N, has been imposed on n and is needed in connection with proofs of asymptotic (with T) properties of the method. (N might increase with T.) In practice such bounds do not seem to be used probably because the bounds needed for validity are much larger than values of n that an experienced investigator would consider reasonable and are needed in the theoretical investigation only to exclude ridiculously large values.

For the case of $C_T = \log T$ a justification has been given by Rissanen on the basis of a minimum description length principle. The idea is to use the model set to record the data in as few bits as possible. The first term (or rather $T/2$ by it) gives a measure of the average number of bits required for an optimal encoding when n is fixed and the maximum likelihood structure, on Gaussian assumptions, is taken to be the true structure. To decode, the model parameters must also be transmitted and $T/2$ by the second term in (3.1), for BIC, measures the number of bits for an optimal encoding of these, to an accuracy determined by that of the method of maximum likelihood. The use of $C_T \equiv 2$ has been justified by Akaike on the basis of a prediction theory, and has been widely used.

The emphasis in this chapter will principally be on the use of rational transfer function systems as approximations to systems of a more general kind. This will be further discussed in the next section. However here some discussion of the case where there is a true rational transfer function system will be given in relation to the use of (3.1). The conditions under which the statements below hold true are essentially (6.1), (6.2), below, plus the finiteness of fourth moments of the $e_j(t)$, but the proofs of the theorems also depend on a condition $\det W(z) \neq 0$, $|z| \geq 1-\delta$, $\delta > 0$. (Compare below (1.1).) This δ may be as small as desired but is prescribed *a priori*.

Now assume there is a true n_0 and $\hat{n}$ minimises (3.1) while, as $T \to \infty$, $C_T/T \to 0$ (which is an insignificant requirement). Then the following holds, where a.s. stands for "almost surely".

(i) If $\liminf_{T \to \infty} C_T/(2\log\log T) > 1$ then $\hat{n} \to n_0$, a.s.

If $\limsup_{T \to \infty} C_T/(2\log\log T) < 1$ then $\hat{n}$ does not converge a.s. to n_0.

(ii) If $\liminf_{T \to \infty} C_T = \infty$ then $\hat{n} \to n_0$ in probability.

If $\limsup_{T \to \infty} C_T < \infty$ then

$$\lim_{\delta \to 0} \lim_{T \to \infty} P\{\hat{n} > n_0\} = 1. \tag{3.2}$$

These results deserve careful interpretation. In the first place (i) should not be interpreted as saying that $C_T = 2\log\log T$ is a good value to use because $2\log\log T$ changes far too slowly with T to be meaningful. At $T = 10$ it is 1.7 and at $T = 1000$ it is 3.9. It is therefore not far from $C_T = 2$ for most values of T met in practice, which is the value for AIC(n). The result (3.2) suggests that AIC(n) is bad because it will always overestimate the McMillan degree. However in practice there will be no true degree and then $\hat{n}$ should increase with T. The question is how fast. Some investigations suggest that $C_T = 2$, i.e. AIC, gives an optimal rate of increase, according to certain criteria.

The result (3.2) deserves further discussion. We give this for the simplest case where $q = 1$, $n_0 = 0$ so that $y(t) = e(t)$. When $n = 1$ is the model then

$$y(t) + ay(t-1) = e(t) + ce(t-1), \ |a| < 1, \ |c| < 1-\delta. \quad (3.3)$$

We indicate why $n = 1$ will be preferred to $n = 0$, the true value, when C_T is uniformly bounded. The choice between the two values will be based on

$$\log \hat{\Omega}_1 + 2C_T/T - \log \hat{\Omega}_0 = -\log(\hat{\Omega}_0/\hat{\Omega}_1) + 2C_T/T$$

so that $n = 1$ will be preferred when $\lambda_T = T \log(\hat{\Omega}_0/\hat{\Omega}_1) > 2C_T$. Consider Fig. 1.

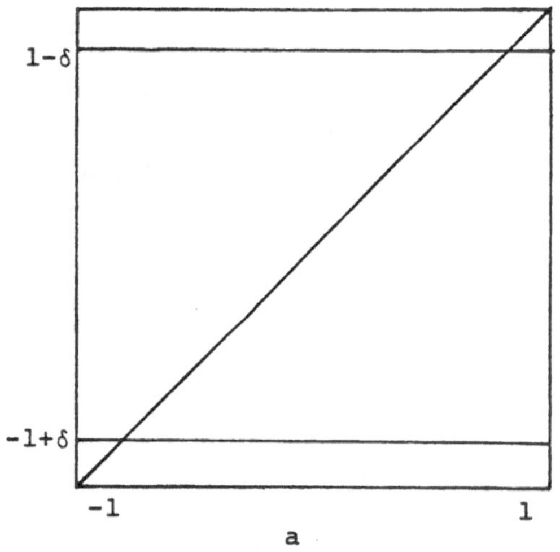

Figure 1.

The region of optimisation for $n = 1$ is that below and above the lines through $\pm(1-\delta)$, excluding the diagonal, but the maximum of the likelihood could be at the boundary. In fact if $\hat{a}$, $\hat{c}$ are the maximum likelihood estimates it may be shown that $(\hat{a} - \hat{c}) \to 0$, so that $(\hat{a},\hat{c})$ moves to the diagonal. Thus λ_T is eventually the maximum of a function defined on the diagonal of Fig. 1. Let us parameterise that diagonal by $\alpha = \log\{(1+a)/(1-a)\}$ so that $|\alpha| < \log\{(2-\delta)/\delta\} = \Delta$, let us say. Then this function, of which λ_T is eventually the maximum value, is $\xi(\alpha)^2$ where $\xi(\alpha)$ is a stationary random function of α, i.e. α takes the place of t

in our previous considerations but varies continuously. $\xi(\alpha)$ has spectral density $\{\cosh \pi\omega\}^{-1}$ $-\infty < \omega < \infty$. It is evident that $\xi(\alpha)$ will, as $\delta \to 0$ so that Δ becomes increasingly large, take its maximum for increasingly large values of α i.e. values of a that approach ± 1. Thus $(\hat{a},\hat{c})$ will approach $(1,1)$ or $(-1,-1)$, which will make optimisation difficult. This result has another interpretation as follows. It is approximately true that $\hat{\Omega}_1$ is the minimum value of

$$\frac{1}{2\pi}\int_{-\pi}^{\pi} \{|d(\omega)|^2/|W(e^{i\omega})|^2\}\, d\omega \tag{3.4}$$

where $W(z) = (z-c)/(z-a)$. (See section 2(i) for $d(\omega)$). We know that that $\hat{c}-\hat{a} \to 0$. If $\hat{c},\hat{a}$ remain bounded away from ± 1 then $|W(e^{i\omega})|^2 \to 1$, uniformly, as $\hat{c}-\hat{a} \to 0$. The method of maximum likelihood attempts to minimise (3.4), so that it seeks to move towards $(\hat{c},\hat{a}) = (1,1)$ or $(-1,-1)$. If $\hat{a}$ goes to ± 1 faster than $\hat{c}$ does, then $|W(e^{i\omega})|^{-2}$ becomes zero at $\pm\pi$ (for $+1$) or 0 (for -1) and thus (3.4) is further reduced (since $|W(e^{i\omega})|^2$ will converge to unity uniformly at other values of ω). Thus $|W(e^{i\omega})|^2$ develops a "notch" at $\pm\pi$ or 0. Where the notch will be and what its precise shape will be is determined by the shape of $|d(\omega)|^2$ near 0, $\pm\pi$. This function is incredibly irregular for T large, so that there will be many values for $\hat{a},\hat{c}$ that will give local minima to (3.4) and the absolute minimum may be very difficult to find. This corresponds to the fact that $\xi(\alpha)^2$ will have many local maxima and minima, which will be compressed into small neighbourhoods of $a = \pm 1$ (for δ small) because of the nature of the function $\alpha = \log\{(1+a)/(1-a)\}$.

The general situation, i.e. that for general n, q, p, is essentially the same. It must be emphasised that (3.2) is very "asymptotic" in that T may need to be very large before it is relevant.

Notes on References. The procedures described in this section were suggested in Akaike (1969), Rissanen (1983). The results in (3.2) and above are in Hannan (1980), (1981), (1984). For the results relating to AIC when there is no true n_0 see Shibata (1980), Hannan and Kavalieris (1980).

4. Rational Transfer Function Approximation

In this section a brief account will be given of some deep theory concerning the approximation of the true structure by approximating to H by a Hankel matrix of finite rank. (Readers less concerned with theory may choose to "skip" this section.) The methods relate mostly to the case where there are no inputs and only that case will be discussed here. The idea is to approximate to $W(z)$ by a $\tilde{W}(z)$ for n finite so that $H-\tilde{H}$ is as small as possible in the Hankel norm, which is the Euclidean norm (or singular value norm) for H as an operator from one Hilbert space to another. To see what is involved put

$$y_t = (y(t)',y(t+1)',\ldots)', \quad e_t = (e(t)',e(t+1)',\ldots)'$$

$$e^t = (e(t-1)',e(t-2)',\ldots)' \quad y^t = (y(t-1)',y(t-2)',\ldots)'.$$

Then, as is easily checked from (1.1),

$$y_{t+1} = He^t + Ke_{t+1}, \quad E(e^t e_{t+1}') = 0, \tag{4.1}$$

where K has $W_{j-k}, j,k=1,2,\ldots$ as the (j,k)th block, $W_j = 0$, $j < 0$. Thus (4.1) describes the dependence of the future on the past. The space on which H operates is therefore endowed with a metric structure given by the covariance matrix of e^t, namely $(I_\infty \otimes \Omega)$. By $(I_\infty \otimes \Omega)$ we mean the Tensor product of the two matrices i.e. a block diagonal matrix with the diagonal blocks being Ω. For the general definition of tensor product see below (5.5). The space to which H operates is endowed with a metric structure given by the covariance matrix of y_{t+1}, namely that with (j,k)th block

$$E\{y(t+j)y(t+k)'\} = \Gamma(k-j) = \Gamma(j-k)'.$$

Since $\Gamma(t)'$ is the t'th Fourier coefficient matrix of $f(-\omega)$ we consider the canonical factorisation of this, as for $f(\omega)$ in (1.2). Let this factorisation be $f(-\omega) = (2\pi)^{-1}\tilde{W}(e^{-i\omega})\tilde{\Omega}\tilde{W}(e^{-i\omega})^*$. (This notation is in agreement with that in section 2 because $f(-\omega)$ is the spectrum of the time reversed process.) Here $\tilde{W}(z) = \Sigma\,\tilde{W}_j z^{-j}$

and $\det \tilde{W} \neq 0$, $|z| \geqslant 1$. Let $\mathcal{W}$ have $\tilde{W}(k-j)$ as the (j,k)th block, $\tilde{W}(j) = 0$, $j < 0$. Then

$$S = (I_\infty \otimes \tilde{\Omega}^{-\frac{1}{2}})\, \mathcal{W}^{-1} \mathcal{H} (I_\infty \otimes \Omega^{\frac{1}{2}}) \tag{4.2}$$

operates from ℓ_2 to ℓ_2, where ℓ_2 is the space of all sequences $a_1, a_2, \ldots$ with $\Sigma|a_j|^2 < \infty$ and with the inner product $(a,b) = \Sigma a_j \bar{b}_j$. Thus it is S whose singular value decomposition is sought. S is also a Hankel matrix because $\mathcal{W}$ is upper triangular and block Toeplitz, so that $\mathcal{W}^{-1}$ is also of that form. It follows then that $\mathcal{W}^{-1}\mathcal{H}$ is of Hankel matrix form. The blocks, S_{j+k-1}, $j,k = 1,2,\ldots,$ in the typical (j,k)th place in S are generated by the matrix function

$$S(z) = \tilde{\Omega}^{-\frac{1}{2}}\, \tilde{W}(z)^{-1}\, W(z^{-1})\, \Omega^{\frac{1}{2}} \tag{4.3}$$

and for $z = \exp i\omega$ it is easily checked that this is a unitary matrix. If $q = 1$ then $f(-\omega) = f(\omega)$ so that $\tilde{\Omega} = \Omega$ and $\tilde{W} = W$. <u>In the scalar case, $q = 1$, we shall in future use lower case letters</u>. Thus we write $w(z)$. In this case therefore (4.3) becomes

$$s(z) = w(z^{-1})/w(z)$$

which is obviously of modulus 1 for $z = \exp i\omega$ since $w(z)$ has real coefficients. Of course $S(z)$, unlike $W(z)$, is not analytic for $|z| \leq 1$. However only the coefficients of the analytic part, i.e. the coefficients of z^j, $j > 0$, occur in S. The singular value decomposition of S is of the form (assuming that operator to be appropriate, e.g. compact)

$$S = \sum_1^\infty \rho_j \eta_j \xi_j^*, \quad \eta_j^* \eta_k = \xi_j^* \xi_k = \delta_{jk} \quad \rho_1 \geq \rho_2 \geq \ldots \geq 0.$$

Introduce the new random variables

$$u_j = \eta_j^* (I_\infty \otimes \tilde{\Omega}^{\frac{1}{2}})\, \mathcal{W}^{-1}\, y_{t+1} \quad ,$$

$$x_j = \xi_j^* (I_\infty \otimes \Omega^{-\frac{1}{2}})\, e^t.$$

Then

$$E(u_j \bar{u}_k) = E(x_j \bar{x}_k) = \delta_{jk}; \quad E(u_j \bar{x}_k) = \delta_{jk}\rho_j.$$

The u_j, x_j might be called "discriminant functions" since they occur in the classical theory of statistical canonical correlation analysis as functions that are used to classify individuals. The

ρ_j themselves would be called "canonical correlations". Since e^s, $s \leq t$, spans the same space as do the y^s, $s \leq t$, the same canonical correlations and the same u_j (but not the same x_j) would be obtained if S were considered as an operator from a space with the metric structure determined by the covariance matrix of y^t. Once the singular value decomposition of δ is known it is possible uniquely (at least for $q = 1$) to determine the best Hankel norm approximation to H and equivalent to $W(z)$, for given n. We shall not enter further into that here since the virtues of such an approximation are by no means evident in a statistical context nor are the effects of having to estimate the ρ_j, u_j, x_j. However these ideas have been used by Akaike in a way that we shall briefly survey, after first introducing a canonical representation (1.6), for the ARMA case.

It will be recalled that such a canonical form is attained if H_0 is chosen as constituted by the first n linearly independent rows of H. Call $r(v,j)$ the j'th row, $j = 1,\ldots,q$, in the v'th block, $v=1,2,\ldots$. Then such a set of rows is always of the form

$$r(v,j), \quad v = 1,\ldots,n_j; \quad j = 1,\ldots,q; \quad \Sigma n_j = n. \qquad (4.4)$$

The n_j are known as the Kronecker indices. They uniquely determine these first linearly independent rows of H_0. There is a corresponding unique factorisation of $W(z) = A(z^{-1})^{-1}C(z^{-1})$, where $A(z^{-1}) = Z\tilde{A}(z)$, $C(z) = Z\tilde{C}(z)$ and Z is diagonal with z^{-n_j} in the j'th place in the diagonal. A, C are matrices of polynomials with $\tilde{A}$ having diagonal elements of degree n_j which are monic, i.e. have unity as the coefficient of z^{n_j}. Putting $\tilde{E} = \tilde{C} - \tilde{A}$ the decomposition is uniquely defined by the inequalities on degrees

$$\deg \tilde{a}_{ij} < \deg \tilde{a}_{jj},\ j \neq i; \quad \deg \tilde{a}_{ij} \leq \deg \tilde{a}_{ii},\ j \leq i;$$

$$\deg \tilde{a}_{ij} < \deg \tilde{a}_{ii},\ j > i$$

$$\deg \tilde{e}_{ij} < \deg \tilde{a}_{ii}, \quad i,j = 1,2,\ldots,q.$$

Akaike's method leads to estimates of the n_j and of $\tilde{A}$. Put $\tilde{y}^t = (y(t)', y(t-1)',\ldots,y(t-h)')'$ where h might be chosen by fitting an autoregression and determining h as the order minimising BIC or AIC. Put, for $\ell = 0,1,\ldots;\ m = 0,1,\ldots,q-1$,

$$y_{\ell m}(t)' = (y(t+1)', y(t+2)', \ldots, y(t+\ell)', y_1(t+\ell+1), \ldots, y_m(t+\ell+1))'.$$

If the smallest n_j is for $j = m$ and $n_m = \ell$ then row $r(\ell+1,m)$ (see (4.4)) is linearly dependent on earlier rows of H and correspondingly (see (4.1)) there will be some linear function of $y_{\ell,m}(t)$ that is orthogonal to the past, while this will not be true for $\ell_1 < \ell$ or for $\ell_1 = \ell$, $m_1 < m$. To judge when this is so we consider the solutions of

$$\hat{\eta}_j'[\hat{\rho}_j I_{\ell q+m} - \hat{B}_{\ell,m}\hat{B}_{\ell,m}'], \quad \hat{\rho}_1 \geq \hat{\rho}_2 \geq \ldots \geq \hat{\rho}_{\ell q+m}$$

$$\hat{B}_{\ell,m} = \{\frac{1}{T}\Sigma y_{\ell,m}(t) y_{\ell,m}(t)'\}^{-\frac{1}{2}} \frac{1}{T}\Sigma y_{\ell,m}(t)(\tilde{y}^t)'\{\frac{1}{T}\Sigma\tilde{y}^t(\tilde{y}^t)'\}^{-\frac{1}{2}}$$

where the summations are over $h+1 \leq t \leq T-\ell-1$. It is assumed that $\ell q+m \leq hq$. The $\hat{\rho}_j$ are the canonical correlations between the $y_{\ell,m}(t)$ and $\tilde{y}^t$. Successively examining these canonical correlations (ordering (ℓ,m) in dictionary order, first according to ℓ and then m) we stop when, for the first time

$$-(T-\nu_{\ell,m})\log(1-\hat{\rho}^2_{\ell q+m}) - \nu_{\ell,m} > 0; \qquad \nu_{\ell,m} = q(h-\ell)-m+1.$$

If this happens at $\ell(1)$, $m(1)$ then $\hat{n}_{m(1)}$ is put at $\ell(1)$. Now eliminate $y_{m(1)}(t+\ell(1)+j)$, $j > 0$, from all future $y_{\ell,m}$ and continue, always taking $\nu_{\ell,m}$ as qh-dim $y_{\ell,m}(t) + 1$. Once $\hat{n}_{m(1)}$ is determined we eliminate $y_{m(2)}(t+\ell(2)+j)$, $j > 0$, from future $y_{\ell,m}(t)$ and continue and so on. In this way all $\hat{n}_k$ are determined and with each will be associated an $\hat{\eta}(k)$, which is the $\hat{\eta}_j$ for the smallest $\hat{\rho}_j$ at the step when $\hat{n}_k$ was determined. $\hat{\eta}_j$ is determined only up to a scalar factor and that is fixed in $\hat{\eta}(k)$ by making the last element unity. Now $\hat{\eta}(k)$ determines the k'th row of the estimate of $\tilde{A}(z)$ so that the element of $\hat{\eta}(k)$ corresponding to $y_j(t+v)$ in $y_{\ell,m}(t)$, for ℓ,m at the values where $\hat{n}_k$ was determined, is the coefficient of z^{v-1} in the estimate of $\tilde{a}_{k,j}(z)$. Thus at the end of the calculation the $\hat{n}_k$ and estimate $\hat{A}$ of $A(z)$, in canonical form corresponding to the Kronecker indices, are available. It is then necessary to estimate $C(z)$. This would be done by forming $\hat{A}(z^{-1})y(t)$ and using the calculated autocovariances of this to estimate those of $C(z^{-1})e(t)$. Then an estimate of the spectrum will be obtained and factored to find an estimate of $C(z)$. Since $A(0) = C(0)$ and the row degrees of $C(z)$ are prescribed by the degree inequalities this would have to be done carefully and would not be a trivial calculation for $q > 1$. In any case these estimates of A(z), C(z) are inefficient but could be used to initiate a minimisation of (2.5), in the form for (1.5) corresponding to the canonical choice of H_0 and the $\hat{n}_k$. We do not proceed further with the description because there are problems with the method. It is, so far, restricted to the ARMA case. The $\hat{n}_k$ are determined in an inefficient estimation procedure and no later adjustment of them has been suggested. However the method is of interest because of its association with the theory of the first part of this section.

Notes on References. Adamyan, Arov and Krein (1971), Glover (1983) and Jewell and Bloomfield (1983) deal with the theory of Hankel norm approximation. Jewell and Bloomfield (1983,a) suggest, for q=1, that the canonical correlations be found directly from $s(z) = W(z)/W(z^{-1})$, which is to be obtained by factoring a spectral estimate. Akaike (1969,a) presents his method. For some estimation of a moving-average model see Hannan (1970).

5. A Gauss-Newton Procedure

(i) First the case q=1 will be discussed because this is important and the calculations are then quite feasible. The idea is to use a Gauss-Newton procedure to approximate to the true ARMAX structure but to include n in the estimation. At each iteration this is to be done recursively. Thus consider

$$\frac{1}{T}\sum_{1}^{T} e_\tau(t)^2, \quad e_\tau(t) = c_\tau(z^{-1})^{-1}\{a_\tau(z^{-1})y(t)-b_\tau(z^{-1})u(t)\}. \qquad (5.1)$$

Here a_τ, b_τ, c_τ are the transfer functions, for given n, in the ARMAX model for q=1 and τ is the vector of system parameters i.e. the 3n freely varying coefficients in a_τ, b_τ, c_τ. Here, again, we use lower case letters for the scalar case. Note that b_τ is, in general, a row vector since we do not require p=1. The $e_\tau(t)$ are functions only of $w_\tau(z)^{-1} = c_\tau(z^{-1})^{-1}a_\tau(z^{-1})$ and $w_\tau(z)^{-1}\ell_\tau(z) = c_\tau(z^{-1})^{-1}b_\tau(z^{-1})$. The procedure is to linearise these functions about a previous estimate, which reduces the minimisation of (5.1) to a linear problem. As has been said the procedure is Gauss-Newton but includes n in the optimisation. It is necessary to obtain a first estimate from which to commence the iteration. This is done by taking $c_\tau \equiv 1$ and choosing a_τ, b_τ by regression autoregression. We go on to describe the algorithm.

Step 0. Put

$$v(t) = \begin{pmatrix} y(t) \\ -u(t) \end{pmatrix}, \quad t = 1,\ldots,T$$

and use the Levinson-Whittle algorithm. Let $\hat{\sigma}_n^2$ be the top left hand element of S_n. Choose $\hat{n}$ to minimise

$$\log \hat{\sigma}_n^2 + n(p+1)\log T/T.$$

Let the first row of $F_{\hat{n},j}$ be called $(\hat{a}_j, \hat{b}_j')$ where $\hat{a}_j$ is scalar and $\hat{b}_j$ has p elements. Then $\hat{a}_j$ is the j'th coefficient in $\hat{a}(z)$ and $\hat{b}_j'$ is the j'th coefficient vector in $\hat{b}(z)$.

The basic algorithm is now given by step 1 which is repeated until convergence. To commence step 1 one needs estimates $\hat{n}$, $\hat{a}$, $\hat{b}$, $\hat{c}$. These will initially come from step 0, with $\hat{c} \equiv 1$.

Step 1. Define $\hat{e}(t)$, $\hat{\eta}(t)$, $\hat{\xi}(t)$, $\hat{\zeta}(t)$ by

$$\hat{c}\hat{e}(t)=\hat{a}y(t),\ \hat{c}\hat{\eta}(t)=y(t),\ \hat{c}\hat{\xi}(t)=\hat{e}(t),\ \hat{c}\hat{\zeta}(t)=u(t),$$
$$y(t)=\hat{\eta}(t)=\hat{\xi}(t)=\hat{\zeta}(t)=\hat{e}(t)=0, \quad t \leqslant 0.$$

Put

$$v(t) = \begin{pmatrix} \hat{\eta}(t) \\ -\hat{\zeta}(t) \\ -\hat{\xi}(t) \end{pmatrix}, \quad t = 1,2,\ldots,T$$

and use the Levinson-Whittle recursion to generate $F_{n,j}$, $\tilde{F}_{n,j}$, S_n, $\tilde{S}_n$.
Put

$$\hat{r}_n(t) = \sum_0^n \tilde{F}_{n,j} v(t-n+j)$$

$$\begin{pmatrix} \hat{a}_{n,n}^{(1)} \\ \hat{b}_{n,n}^{(1)} \\ \hat{c}_{n,n}^{(1)} \end{pmatrix} = \tilde{S}_{n-1}^{-1} \frac{1}{T} \sum_1^T \{\hat{\eta}(t)-\hat{\xi}(t)+\hat{e}(t)\}\ \hat{r}_{n-1}(t-1) \tag{5.2}$$

$$\begin{pmatrix} \hat{a}_{n,j}^{(1)} \\ \hat{b}_{n,j}^{(1)} \\ \hat{c}_{n,j}^{(1)} \end{pmatrix} = \begin{pmatrix} \hat{a}_{n-1,j}^{(1)} \\ \hat{b}_{n-1,j}^{(1)} \\ \hat{c}_{n-1,j}^{(1)} \end{pmatrix} + \tilde{F}_{n-1}(n-j)' \begin{pmatrix} \hat{a}_{n,n}^{(1)} \\ \hat{b}_{n,n}^{(1)} \\ \hat{c}_{n,n}^{(1)} \end{pmatrix}, \quad j = 1,2,\ldots,n-1.$$

$$\hat{a}_{n,0}^{(1)} \equiv \hat{c}_{n,0}^{(1)} \equiv 1, \quad \hat{b}_{n,0}^{(1)} \equiv 0.$$

$$\hat{\sigma}_n^2 = \hat{\sigma}_{n-1}^2 - (\hat{a}_{n,n}^{(1)}, \hat{b}_{n,n}^{(1)\prime}, \hat{c}_{n,n}^{(1)})\ \tilde{S}_{n-1} \begin{pmatrix} \hat{a}_{n,n}^{(1)} \\ \hat{b}_{n,n}^{(1)} \\ \hat{c}_{n,n}^{(1)} \end{pmatrix}$$

$$\hat{\sigma}_0^2 = \frac{1}{T} \sum_1^T \{\hat{\eta}(t) - \hat{\xi}(t) + \hat{e}(t)\}^2$$

Choose $\hat{n}^{(1)}$ to minimise

$$\log \hat{\sigma}_n^2 + n(p+2) \log T/T$$

and set $\hat{n} = \hat{n}^{(1)}$, $\hat{a}(t) = \Sigma \hat{a}_j z^j$, $\hat{c}(z) = \Sigma \hat{c}_j z^i$, $\hat{b}(z) = \Sigma \hat{b}_j' z^j$

$$\hat{a}_j = \hat{a}_{\hat{n},j}^{(1)}, \quad \hat{b}_j = \hat{b}_{\hat{n},j}^{(1)}, \quad \hat{c}_j = \hat{c}_{\hat{n},j}^{(1)},$$

and proceed to repeat step 1.

Cease the iteration when the minimised value of $\log \sigma_n^2 + n(p+2) \log T/T$ stabilises.

We make a number of remarks.

1. If there is a true rational transfer function system then this algorithm will provide estimates that converge to the true values as T increases. Under fairly general conditions on the $e(t)$ (see section 7) $T^{\frac{1}{2}}(\tau-\tau_0)$, τ_0 being the true vector of system parameters, will be asymptotically normal. The covariance matrix could be estimated as

$$\sigma^2 \{\frac{1}{T} \sum_t \begin{pmatrix} v(t) \\ v(t-1) \\ \vdots \\ v(t-\hat{n}) \end{pmatrix} (v(t)', v(t-1)', \ldots, v(t-\hat{n}))\}^{-1}$$

where $\hat{\sigma}^2$ is $\hat{\sigma}_{\hat{n}}^2$ at the last iteration and $v(t)$ is that vector used at the last iteration. A more efficient estimate of the covariance matrix would be obtained via the $\hat{a}_j$, $\hat{b}_j$, $\hat{c}_j$, $j=1,\ldots,\hat{n}$, $\hat{\sigma}^2$, at the last iteration but we omit the formulae for brevity.

2. It is not fully apparent that it is best to use BIC and some would argue that in a situation where it is unreasonable to believe in a true rational transfer function system then AIC should be used. Much depends upon the end purpose of the analysis.

3. Though the Levinson-Whittle recursion has been used above this could be replaced by other recursive calculations of the form discussed briefly in sub-section 2(ii). An alternative would be finally to effect one iteration of an algorithm to optimise (2.5) at $n=\hat{n}$ and initiating with $\hat{a}_j$, $\hat{b}_j$, $\hat{c}_j$, $j=1,\ldots,\hat{n},\hat{\sigma}^2$.

4. One problem with the algorithm is that unless $\hat{c}(z) \neq 0$, $|z| \leqslant 1$, then it will fail at step 1 since $\hat{e}(t)$ etc will grow exponentially with t. What should be done is to reflect those zeros of $\hat{c}(z)$, that are inside the unit circle, in the unit circle. This may be

done as follows. Form

$$\hat{\sigma}^2 \sum_0^{\hat{n}} \hat{c}_{\hat{n},j} z^{-j} \sum_0^{\hat{n}} \hat{c}_{\hat{n},j} z^j$$

which will be factored as

$$\tilde{\sigma}^2 \sum_0^{\hat{n}} \tilde{c}_{\hat{n},j} z^{-j} \sum_0^{\hat{n}} \tilde{c}_{\hat{n},j} z^j$$

where now $\Sigma \tilde{c}_{\hat{n},j} z^j \neq 0,\ |z| \leq 1$. To achieve this form

$$\rho(j) = \sum_{k=0}^{\hat{n}-j} \hat{c}_{\hat{n},j+k} \hat{c}_{\hat{n},k} \Big/ \sum_0^{\hat{n}} \hat{c}^2_{\hat{n},j}$$

and put

$$\delta_{j+1} = \frac{1}{2}\delta_j + \frac{1}{2}\hat{\delta}_j, \qquad \delta_0' = (1,0,0,\ \ldots 0)$$

$$\hat{\delta}_j' = 2(1,\rho(1),\ldots,\rho(\hat{n}))G^{-1}(\delta_j)$$

$$G(\delta_j) = \begin{bmatrix} \delta(0) & \delta(1) & \ldots & \delta(\hat{n}) \\ \delta(1) & \delta(2) & \ldots & 0 \\ \vdots & & \vdots & \vdots \\ \delta(\hat{n}) & 0 & \ldots & 0 \end{bmatrix} + \begin{bmatrix} \delta(0) & \delta(1) & \ldots & \delta(\hat{n}) \\ 0 & \delta(0) & \ldots & \delta(\hat{n}-1) \\ \vdots & \vdots & \vdots & \vdots \\ 0 & 0 & \ldots & \delta(0) \end{bmatrix}$$

Here $\delta(k)$ is the k'th element of δ_j. If $\tilde{\delta}$ is the limit of this sequence, δ_j, then

$$\tilde{c}_{\hat{n},j} = \tilde{\delta}(j)/\tilde{\delta}(0).$$

Now $\tilde{c}_{\hat{n},j}$ is used in place of $\hat{c}_{\hat{n},j}$ at step 1.

Instead of performing this calculation before step 1, on each occasion, it may by computationally cheaper to check the location of the zeros of $\hat{c}(z)$ (by a Schur-Cohn algorithm) and use the above algorithm only when the test shows that there are zeros inside $|z| = 1$.

5. In some applications it may be felt that economy in the use of parameters may be effected by allowing the degrees of $a(z)$, $b(z)$, $c(z)$ to differ, for example by taking the degree of $c(z)$ lower. In principle, at each iteration of steps 1 to 4, one can allow these degrees to vary arbitrarily but this will be much more computationally costly. An alternative would be to proceed by eliminating terms only after the last iteration, always using BIC to determine whether this elimination has been worthwhile. We omit details.

6. At the first iteration of step 1 then $\hat{c}(z) \equiv 1$ and hence

$\hat{\eta}(t) = y(t)$, $\hat{\xi}(t) = \hat{e}(t)$, $\hat{\zeta}(t) = u(t)$. Computationally more efficient algorithms have been given for this case, when $u(t)$ does not occur, which exploit the fact that $\hat{e}(t)$ is Toeplitz orthogonal to $y(t-j)$, $j=1,\ldots,\hat{n}$, and that $\hat{e}(t-j)$, $\hat{e}(t-k)$ are approximately orthogonal and could be treated as such. It is possible that a more efficient implementation may also be found at later iterations. It may be mentioned that in confining summations to $t=1,2,\ldots,T$ we have treated $y(t)$, $u(t)$ as zero for $t \leq 0$, but have avoided that assumption for $t > T$. This seems preferable.

7. At step 1 from the first iteration (i.e. repetition) of the procedure one may replace $\hat{\eta}(t)-\hat{\xi}(t)+\hat{e}(t)$ by $\hat{e}(t)$ in (5.2) for $n \geq \hat{n}$. This is because $\hat{c}(z^{-1})\hat{\eta}(t)$ implies that $\hat{a}(z^{-1})\hat{\eta}(t) = \hat{e}(t)$ so that $\hat{a}(z^{-1})\hat{\eta}(t)-\hat{c}\hat{\xi}(t)=0$ and $\hat{\eta}(t)-\hat{\xi}(t)$ is, for $n \geq \hat{n}$ a linear combination of the variables in the regression. When this is done the $\hat{a}^{(1)}_{n,j}$, $\hat{c}^{(1)}_{n,j}$ must be regarded as adjustments to the previous $\hat{a}_{\hat{n},j}$, $\hat{c}_{\hat{n},j}$ i.e. must be added to these.

(ii) Now consider the vector case, which is more elaborate. First return to the set, $M(n)$, of all systems, (1.3), of McMillan degree n, for given Ω. (We fix Ω for the moment only, because it is the system parameters that need discussion.) $M(n)$ is equivalently the set of all Hankel matrices H of rank n (for $W(z)$ obeying the requirements below (1.1)) and of all pairs of transfer functions $W(z)$, $L(z)$. It may be conceptualised as a smooth surface of dimension $n(2q+p)$ and, technically, is an analytic manifold. A reasonable approach to estimating a system would therefore be to determine n and then the appropriate point on $M(n)$ and this is what was done for $q=1$. For $q > 1$ this is however a problem because $M(n)$ cannot then be covered by one neighbourhood that may be mapped homeomorphically into Euclidean space. $M(n)$ is the union of all systems whose Kronecker indices sum to n and hence an alternative to the consideration of $M(n)$ is the determination of the Kronecker indices, as was the technique used in section 4. There is, however, something very arbitrary in the decomposition of $M(n)$ into sets corresponding to different partitions of n as a sum of q integers and the effort required for an efficient procedure to discover these is fairly considerable. Amongst the set of Kronecker indices summing to n there is one special set, namely those which, for $n = qh + m$, $0 \leq m < q$, are of the form $n_1 = n_2 = \ldots n_m = h+1$, $n_{m+1} = \ldots = n_q = h$. Then the first n linearly independent rows in H are just the first n rows. If

U(n) is the subset of M(n) for which these rows are linearly independent then U(n) is open and dense in M(n). Thus little or nothing is lost in restricting attention to U(n). It is most unlikely that the maximum of the likelihood will be found off U(n) in M(n). (However U(n) would provide a bad coordinate system in which to work if the maximum was near the edge.) We describe U(n) in another way by giving a unique description of A(z), B(z), C(z) in $W(z) = A(z^{-1})^{-1}C(z^{-1})$, $L(z) = A(z^{-1})^{-1}B(z^{-1})$ for a system in U(n). We do this by describing the coefficient matrices $A_{n,j}$, $B_{n,j}$, $C_{n,j}$ in A(z), B(z), C(z). These will be depicted below with a star indicating a freely varying submatrix of elements. All partitions are after the m'th row or column

$$A_{n,0} = C_{n,0} = \begin{bmatrix} I_m & 0 \\ * & I_{q-m} \end{bmatrix}, \quad B_{n,0} = 0, \quad A_{n,1} = \begin{bmatrix} * & 0 \\ * & * \end{bmatrix},$$
$$A_{n,h+1},\ B_{n,h+1},\ C_{n,h+1} = \begin{bmatrix} * \\ 0 \end{bmatrix}. \tag{5.3}$$

All other $A_{n,j}$, $B_{n,j}$, $C_{n,j}$, $j \leq h+1$, are unrestricted. We do not mean that $A_{n,h+1}$, $B_{n,h+1}$, $C_{n,h+1}$ are equal. The vector τ of system parameters coordinatising U(n) is of dimension n(2q+p) and is made up of the freely varying elements in the coefficient matrices. We now go on to describe how to estimate n, τ and Ω. We do this by a series of steps that are related to those for q=1 but are more complicated. Steps 0 to 1 are not repeated. Only step 2 is iterated. Always the output from the previous step is the input to the next so we do not indicate those by a special notation i.e. we do not for example write $\hat{A}_j^{(1)}$ for the $\hat{A}_j$ matrix found at step 1 since it is clear which $\hat{A}_j$ is used at step 2 i.e. that from step 1 and not step 0. Also we shall now index the stages in the Levinson-Whittle recursion by h, rather than n as before.

Step 0. Put

$$v(t) = \begin{pmatrix} y(t) \\ -u(t) \end{pmatrix}, \quad t = 1,\ldots,T$$

and use the Levinson_Whittle recursion. Let $\hat{\Omega}_n$ be the top left hand q x q submatrix of S_h, h = 0,1,2,..., where n = qh, and choose $\hat{h}$, i.e. $\hat{n}$, to minimise

$$\log\det\hat{\Omega}_n + n(q+p)\log T/T, \quad n = hq,\ h = 0,1,\ldots\ .$$

Let the first block of q rows in $F_{h,j}$ be called $[\hat{A}_j, \hat{B}_j]$ and let $\hat{\Omega}_n$ be called $\hat{\Omega}$. Then $\hat{A}_j$, $\hat{B}_j$ $j = 1,\ldots,h$ are the coefficient matrices in $\hat{A}(z)$, $\hat{B}(z)$, with $\hat{A}_0 = I_q$, $\hat{C}(z) \equiv I_q$ and $\hat{n} = q\hat{h}$.

Step 1. Put

$$\hat{e}(t) = \sum_{0}^{\hat{h}} \hat{A}_j y(t-j) - \sum_{1}^{\hat{h}} \hat{B}_j u(t-j)$$

and

$$v(t) = \begin{pmatrix} y(t) \\ -u(t) \\ -\hat{e}(t) \end{pmatrix}$$

and use the Levinson-Whittle algorithm. Again the top left hand element of S_h is called $\hat{\Omega}_n$, $n = hq$ and we choose $\hat{h}$ i.e. $\hat{n}$ to minimise

$$\log\det\hat{\Omega}_n + n(2q+p)\log T/T, \quad n = hq, \quad h=0,1,2,\ldots .$$

Now $[\hat{A}_j, \hat{B}_j, \hat{C}_j]$ are the top q rows in $F_{\hat{h},j}$ and provide the coefficient matrices in $\hat{A}(z)$, $\hat{B}(z)$, $\hat{C}(z)$, with $\hat{A}_0 = \hat{B}_0 = I_q$ for $\hat{n} = \hat{h}q$.

Now m has to be determined, in $n = \ell q+m$, $0 \leq m < q$. We choose $\hat{\ell} = \hat{h}-1$ and need only compute for $m = 1,2,\ldots,q-1$, since $m=0$ corresponds to the case $\hat{h}-1$ at step 1, to which $\hat{h}$ was preferred by the criterion. If there is a true rational transfer function system then for large enough T at step 1 $\hat{h}$ will be greater than or equal to the true ℓ, which explains our procedure. We consider $m = q$ but the calculations then are already done at step 1.

The problem is to insert zero elements in the appropriate places in (5.3) and the elements indicated by a star in $A_{n,0}$, $C_{n,0}$. This can be done computationally cheaply using the calculations at step 1 but the details are too complicated to be described here (see the references). It is simpler to describe them as a regression procedure, one for each value of m. (It is unlikely that the algorithm will be used for $q > 5$ and then only 4 or fewer values of m need be taken.) The regression is of a vector variable on $n = (\hat{h}-1)q + m$ other variables, but is carried out row by row so that we describe the calculation for a typical row, j, $j = 1,\ldots,q$. If $1 \leq j \leq m$ we regress $y_j(t)$ on the following variables

(1) $-y_k(t-i)$, $k=1,\ldots,q$; where $i = 1,\ldots,\hat{h}$ for $k \leq m$ and $i = 2,\ldots,\hat{h}$ for $k = m+1,\ldots,q$.

(2) $u_k(t-i)$, $k = 1,\dots,q$; $i = 1,\dots,\hat{h}$

(3) $\hat{e}_k(t-i)$, $k = 1,\dots,q$; $i = 1,\dots,\hat{h}$

where

$$\sum_0^{\hat{h}} \hat{C}_j \hat{e}(t-j) = \sum_0^{\hat{h}} \hat{A}_j y(t-j) - \sum_1^{\hat{h}} \hat{B}_j u(t-j),$$

$$y(t) = u(t) = \hat{e}(t) = 0, \quad t \leqslant 0.$$

For $m < j \leqslant q$ we regress $y_j(t)$ on

(1) $-y_k(t-i)$, $k = 1,\dots,q$; $i = 1,\dots,\hat{h}-1$.

(2) $-(y_k(t) - \hat{e}_k(t))$, $k = m+1,\dots,q$.

(3) $u_k(t-i)$, $\hat{e}_k(t-i)$, $k = 1,\dots,q$, $i = 1,\dots,\hat{h}-1$.

The coefficient of $-y_k(t-i)$ or $-(y_k(t) - \hat{e}_k(t))$ is the j'th regression estimate, $a_{j,k}(i)$, in $A(z)$ and similarly for $u_k(t-i)$, $\hat{e}_k(t-i)$ in relation to $B(z)$, $C(z)$. The matrix $\hat{\Omega}_n$, $n=(\hat{h}-1)q+m$, is estimated as T^{-1} by the sums of squares and cross products of the residuals from the q regressions. Now $\hat{n}$ is chosen by choosing $\hat{m}$, to minimise

$$\log\det \hat{\Omega}_n + n(2q+p)\log T/T, \quad n = q(\hat{h}-1) + m, \quad m = 1,\dots,q.$$

(For $m = q$ the left side is just the minimised expression at the end of step 1).

Now we have an $\hat{n}$, $\hat{A}(z)$, $\hat{B}(z)$, $\hat{C}(z)$ with the latter of the form indicated by (5.3) for $n = \hat{n}$.

As was said above steps 0, 1 are not repeated. Step 2 may be but often no repetition will be necessary, or at most one.

Step 2. Form matrices $\eta(t)$, $\xi(t)$, $\zeta(t)$, of q rows and, respectively q^2, q^2 and qp columns by solving

$$\sum_0^{\hat{h}} \hat{C}_j[\eta(t-j), \zeta(t-j), \xi(t-j)] = (y(t)', u(t)', \hat{e}(t)') \otimes I_q, \tag{5.4}$$

$$(y(t)', u(t)', e(t)') = 0, \quad t \leq 0.$$

Here $\hat{e}(t)$ is obtained from

$$\sum_0^{\hat{h}} \hat{C}_j \hat{e}(t-j) = -\sum_1^{\hat{h}} \hat{B}_j u(t-j) + \sum_0^{\hat{h}} \hat{A}_j y(t-j) \tag{5.5}$$

with the usual initial conditions. By $X \otimes Y$ we mean the tensor product wherein a typical block is $x_{ij}Y$, $i = 1,\dots,a$; $j = 1,\dots,b$

where X is $a \times b$. Of course in (5.4) X is $1 \times (2q+p)$ and all blocks are a scalar multiple of I_q. Thus the $i+q(j-1)$th column of $\eta(t)$, for example, is $\hat{C}(z^{-1})^{-1}E_{ij}y(t)$, where E_{ij} consists of zeros save for a unit in the (i,j)th place.

Put

$$\hat{\Gamma}_v(j) = \frac{1}{T}\Sigma \begin{bmatrix} \eta(t)' \\ -\zeta(t)' \\ -\xi(t)' \end{bmatrix} \hat{\Omega}^{-1}[\eta(t+j), -\zeta(t+j), -\xi(t+j)]. \qquad (5.6)$$

This matrix is of dimension $q(2q + p)$. It is to be the $\hat{\Gamma}_v(j)$ that is the input to the Levinson-Whittle recursion which is to be carried out. It is, thus, $q(2q + p)$ that determines the computational effort. For $q = p = 5$ this is 75, which already would be a rather large scale implementation of the Levinson-Whittle recursion. In cases where q is larger it may be necessary to use some other expedient and we discuss this in remarks below.

Let $\hat{\eta}(t)$ be the vector obtained by adding columns numbered $i + q(i - 1)$, $i = 1,\ldots,q$ in $\eta(t)$ and similarly for $\hat{\xi}(t)$, $\hat{\zeta}(t)$ in relation to $\xi(t)$, $u(t)$. (It is $\hat{\eta}(t)$, $\hat{\xi}(t)$, $\hat{\zeta}(t)$ that correspond most closely to the quantities defined for $q=1$.) Thus

$$\sum_0^{\hat{h}} \hat{C}_j\hat{\eta}(t-j) = y(t), \quad \sum_0^{h} \hat{C}_j\hat{\xi}(t-j) = \hat{e}(t), \quad \sum_0^{\hat{h}} \hat{C}_j\hat{\zeta}(t-j)=u(t).$$

Now form, for each h value considered in the Levinson-Whittle recursion with (5.6),

$$\hat{\tau}_{h,h} = \tilde{S}_{h-1}^{-1} \sum_0^{h-1} \tilde{F}_{h-1,j} \frac{1}{T}\Sigma[\eta(t-h+j), -\zeta(t-h+j), -\xi(t-h+j)]' \{\hat{\eta}(t) - \hat{\xi}(t) + \hat{e}(t)\}.$$

Here $\hat{e}(t)$ is as from (5.5). This vector, $\hat{\tau}_{h,h}$, is of dimension $q(2q+p)$.

$$\hat{\tau}_{h,j} = \hat{\tau}_{h-1,j} + \tilde{F}_{h-1,h-j}\hat{\tau}_{h,h}, \quad j = 0,1,2,\ldots,h-1.$$

To initiate take $\hat{\tau}_{0,0}$ to have zeros everywhere save for units in the places numbered $i + q(i-1)$, $i = 1,\ldots,q$; $q(q+p) + i + q(i-1)$, $i = 1,\ldots,q$. Now the $\hat{\tau}_{h,j}$ provide estimates of the matrices A_j, B_j, C_j, for $n = hq$. Thus $\hat{A}_{h,j}$ has as estimate of $a_{ik}(j)$ the element in the $i + q(k-1)$'th place in $\hat{\tau}_{h,j}$. $\hat{B}_{h,j}$ has as its (i,k)'th element the element in place $q^2 + i + q(k-1)$ in $\hat{\tau}_{h,j}$ while $\hat{C}_{h,j}$ has as its (i,k)'th element that in the $\{q^2 + qp + i + q(k-1)\}$'th place. Next put,

$$\hat{\Omega}_n = \frac{1}{T} \Sigma \hat{e}_n(t)\hat{e}_n(t)', \quad n = hq,$$

where

$$\sum_0^h \hat{C}_{h,j}\hat{e}_n(t-j) = \sum_0^h \hat{A}_{h,j}y(t-j) - \sum_1^h \hat{B}_{h,j}u(t-j)$$

and choose $\hat{n}$ i.e. $\hat{h}$ so that this minimises

$$\log \det \hat{\Omega}_n + n(2q + p) \log T/T, \quad n = hq. \tag{5.7}$$

Now we seek to estimate m in $n = (\hat{h} - 1)q + m$, $m = 1,2,\ldots,q-1$, as in step 1 of the algorithm. Again it will be easiest to describe this as a regression for each m, though it could be computed using the output from the use of (5.6). Consider $[-\eta(t-j), \zeta(t-j), \xi(t-j)]$ formed at (5.4) from the previous step. As in forming $\hat{\tau}_{h,j}$, the $q(2q+p)$ columns in this matrix are associated with the $q(2q+p)$ elements in $[A(j), B(j), C(j)]$ in the sense that the column numbered $i+q(k-1)$, $i,k=1,\ldots,q$ is associated with $a_{ik}(j)$, the column numbered $q^2+i+q(k-1)$, $i=1,\ldots,q$; $k=1,\ldots,p$ is associated with $b_{ik}(j)$ and the column numbered $q^2+qp+i+q(k-1)$, $i,k=1,\ldots,q$ is associated with $c_{ik}(j)$. Now eliminate all columns for which the corresponding element in (5.3) is prescribed to be null. Thus for $j=1$ columns numbered $i+q(k-1)$, $i=1,\ldots,m$; $k=m+1,\ldots,q$ are eliminated. Call the resulting matrix $X_j(t)$ except that for the matrix $X_0(t)$, in addition, columns numbered $i+q(k-1)$, $q^2+qp+i+q(k-1)$ are added, $i=m+1,\ldots,q$, $k=1,\ldots,m$ (all others having been eliminated) to form a matrix of only $m(q-m)$ columns. Now call $\hat{\tau}_n$ the vector of estimates of system parameters for $n=(\hat{h}-1)q+m$, where these parameters are arranged in dictionary order, first according to lag, j, then according to whether the parameter comes from $A(j)$, $B(j)$ or $C(j)$, then according to column index, k and finally according to row index i. Then

$$\hat{\tau}_n = \{\frac{1}{T} \Sigma X(t)'\hat{\Omega}^{-1}X(t)\}^{-1} \{\frac{1}{T} \Sigma X(t)'\hat{\Omega}^{-1}(\hat{\eta}(t) + \hat{e}(t) - \hat{\xi}(t))\}, \tag{5.8}$$

$$X(t) = [X_0(t), X_1(t), \ldots].$$

We emphasise that $X(t)$, $\hat{\Omega}$, $\hat{\eta}(t)$, $\hat{e}(t)$, $\hat{\xi}(t)$ are all formed using the output from the previous step, which at the first use of step 2 will have been step 1, but later will have been from a previous use of step 2. Only $\hat{h}$ has been determined by previous calculations at this step. The notation $\hat{\tau}_n$ in (5.8) should not be confused with $\hat{\tau}_{h,j}$ earlier. $\hat{\tau}_n$ is made up of many submatrices of the type of $\hat{\tau}_{h,j}$ and is of dimension $n(2q + p)$, $n = (\hat{h}-1)q + m$. We now again put

$$\hat{\Omega}_n = \frac{1}{T}\Sigma\,\hat{e}_n(t)\hat{e}_n(t)', \qquad n = (\hat{h}-1)q + m \tag{5.9}$$

$$\sum_0^h \hat{C}_{n,j}\hat{e}_n(t-j) = \sum_0^{\hat{h}} \hat{A}_{n,j}y(t-j) - \sum_1^{\hat{h}} \hat{B}_{n,j}u(t-j)$$

where $\hat{A}_{n,j}$, $\hat{B}_{n,j}$, $\hat{C}_{n,j}$ have elements obtained $\hat{\tau}_n$ according to the identification discussed before (5.7). We choose $\hat{m}$ to minimise

$$\log\det\hat{\Omega}_n + n(2q + p)\log T/T, \qquad n = (\hat{h}-1)q + m, \quad m = 1,2,\ldots,q. \tag{5.10}$$

Again for $m = q$ the value of this criterion is that which optimised (5.7). Then $\hat{A}_j$, $\hat{B}_j$, $\hat{C}_j$ are finally defined to be the values corresponding to $\hat{m}$, $\hat{n}$ is $(\hat{h}-1)q + \hat{m}$ and $\hat{\Omega}$ is the value of $\hat{\Omega}$ from (5.9) that optimised (5.10).

We may now repeat step 2 commencing from these $\hat{A}_j$, $\hat{B}_j$, $\hat{C}_j$, $\hat{\Omega}$, $\hat{n}$ (which defines the $\hat{h}$ in (5.5)). This completes the description of the algorithm.

Remarks. 1. All of the remarks in relation to the scalar case have analogues here. In relation to the use of an estimate of $C(z)$ at (5.4) again a problem will arise unless $\det\hat{C}(z) \neq 0$, $|z| \leq 1$. Again this can be checked via a Schur-Cohn criterion and if that fails we should factor $\hat{C}(e^{i\omega})\hat{\Omega}\hat{C}(e^{i\omega})^*$ canonically so as to obtain a $\tilde{C}(z)$, which does have $\det\tilde{C}(z) \neq 0$, $|z| \leq 1$. Algorithms for this factorisation are available but we omit details here (see the references).

2. Much of the work involved is in step 2 where the sizes of p and q begin to be important. To reduce the calculation there it would not be unreasonable to do them only at the $\hat{h}$, $\hat{m}$ found at step 1. In any case at repetitions of step 2 the values of $\hat{h}$, $\hat{m}$ from the first use of step 2 could be used. When that is done, once (5.4), (5.5) have been computed we may move straight to (5.8), (5.9) for $\hat{h}$, $\hat{m}$ determined at the previous step. However experience with simulations with rational transfer function generated data shows that the determination of $\hat{n}$ is improved at the first use of step 2 and it may improve again at later iterations of that step.

Notes on References. The algorithms here described were first presented in Hannan and Rissanen (1982), Hannan and Kavalieris (1984). The emphasis there was more on order determination. For the determination of m in step 1 (i.e. $q > 1$) an alternative calculation is given in the second reference. For the structure theory at the beginning of subsection (ii) see Diestler and Hannan (1981), for example. The algorithm in remark 4 in subsection (i) is due to Tunnicliffe Wilson (1969) and its matricial version to Tunnicliffe Wilson (1972).

6. Some Theoretical Considerations

This section will be very brief since theory is not the purpose of this account, nor could such theory be fully presented in the space available here. However there seems to be some virtue in indicating the scope of the theory underlying the methods.

In the first place it is not necessary that linear innovations, $e(t)$, be Gaussian and all of the methods are valid under much more general conditions in the sense that the same theory obtains as if the $e(t)$ were Gaussian. The essential condition is that

$$E\{e(t)|e(t-1), e(t-2),\dots\} = 0. \tag{6.1}$$

This is equivalent to the assertion, for $y(t) - \Sigma L_i u(t-i)$, see (1.3), that the best linear predictor is the best predictor (in the least squares sense) so that the data is, in that sense, generated by a linear system. Asymptotic distributions, if they are to be the same as for the Gaussian case, require additionally that

$$E\{e(t)e(t)'|e(t-1),\dots\} = \Omega. \tag{6.2}$$

For $u(t)$ some regularity conditions of a reasonably general nature are needed but we do not discuss them here.

Of course (6.1), (6.2) will hold if the $e(t)$ are independent, with zero mean vector and finite second moments, but are considerably more general.

7. On-Line Procedures

Here only the case $p = q = 1$ will be considered, though the method easily generalises to $p > 1$. There is a large literature concerning methods for real time, on-line estimation of systems and this has recently been surveyed, as will be indicated in the references. Here attention will be concentrated on an on-line implementation, for $q = 1$, of the algorithm described in section 5. In other words we implement the two steps of this algorithm in an on-line fashion, with the step 1 iterated (i.e. repeated) once. Before describing that let us describe three known on-line procedures. Each is of the form

$$\tau(t) = \tau(t-1) + P(t)x(t)\hat{e}(t), \quad \hat{e}(t) = w(t) - \tau(t)'x(t)$$

$$P(t) = \{\sum_{1}^{T} x(t)x(t)'\}^{-1}$$

$$= P(t-1) - \{1 + x(t)'P(t-1)x(t)\}^{-1}P(t-1)x(t)x(t)'P(t-1).$$

Here $v(t)$ is the "independent variable" in a regression on $x(t)$ and $\tau(t)$ is the estimate at time t of the vector of regression coefficients. In the basic on-line procedures $x(t)$, and probably $w(t)$, must be constructed at time t from data to time t together with the estimate $\tau(t-1)$. In each case we identify τ, $x(t)$, $w(t)$.

(1) RLS = Recursive least squares. This corresponds to step 0.

$$\tau(t)' = (a_1(t), a_2(t), \ldots, a_h(t), b_1(t), b_2(t), \ldots, b_h(t)).$$

$$x(t)' = (-y(t-1), -y(t-2), \ldots, -y(t-h), u(t-1), u(t-2), \ldots, u(t-h))$$
$$w(t) = y(t).$$

(2) AML = Approximate maximum likelihood. This corresponds to the first use of step 1.

$$\tau(t)' = (a_1(t), \ldots, a_n(t), b_1(t), \ldots, b_n(t), c_1(t), \ldots, c_n(t))$$

$$x(t)' = (-y(t-1), \ldots, -y(t-n), u(t-1), \ldots, u(t-n), \hat{e}(t-1), \ldots, \hat{e}(t-n))$$
$$w(t) = y(t).$$

In fact what is most properly called AML uses not $\hat{e}(t-j)$ in $x(t)$ but instead $\hat{\varepsilon}(t-j)$

$$\hat{\varepsilon}(t) = y(t) - \tau(t)x(t).$$

This can be done since at time t the latest value used is $\hat{\varepsilon}(t-1)$ which uses $\tau(t-1)$.

(3) RML = Recursive maximum likelihood. This corresponds to the second use of step 1.

$$\tau(t)' = (a_1(t), \ldots, a_n(t), b_1(t), \ldots, b_n(t), c_1(t), \ldots, c_n(t))$$

$$x(t)' = (-\hat{\eta}(t-1), \ldots, -\hat{\eta}(t-n), \hat{\zeta}(t-1), \ldots, \hat{\zeta}(t-n), \hat{\xi}(t-1), \ldots, \hat{\xi}(t-n))$$

$$w(t) = \hat{\eta}(t) + \hat{e}(t) - \hat{\xi}(t). \qquad (7.1)$$

$$\sum_{0}^{n} C_j(t)x(t-j)' = (-y(t), u(t), \hat{e}(t)), \quad C_0(t) \equiv 1, \qquad (7.2)$$

$$y(t) = u(t) = \hat{e}(t) = 0, \quad t \leqslant 0.$$

As presented above each of these could be used as a procedure independently of the others. Of course n is fixed. It is known that AML may not converge, even if the true system is ARMAX of McMillan degree n, unless

$$2\mathcal{R}(c(e^{i\omega})^{-1} - \tfrac{1}{2}) > 0, \qquad \omega \in [-\pi,\pi],$$

i.e. unless the positive real condition is satisfied. It seems that RML may fail unless the location of the zeros of $C(z)$ is monitored and when these move inside the unit circle then the $c_j(t)$ used in forming $\xi(t)$, $\hat{\zeta}(t)$, $\hat{\eta}(t)$, $\hat{e}(t)$ must be held at fixed values outside of the unit circle until the output vector, $\tau(t)$ corresponds to a stable $c_j(t)$ set, i.e. a set with zeros outside of the circle. For these reasons it has been suggested that the algorithms be run in parallel, with the $\hat{e}(t-j)$ for AML provided by the $\hat{e}(t)$ from RLS and with the $\hat{e}(t)$ in (7.1), (7.2) for RML being the $\hat{e}(t)$ from AML. The value of h in RLS would in general be much larger than n in AML, RML where the n is the assumed true order. A common choice would be $h = 2n$, but this is arbitrary.

One main reason for on-line calculation is to allow the estimates to adapt to an evolving mechanism generating the data. In that case one should also be "forgetting" the remote past since that will be irrelevant to the estimation problem. Thus a "forgetting factor" $\ell_t(s)$ is included that multiplies $w(s)$ and $x(s)$ in the calculations at time t. If

$$\ell_t(s) = \prod_{u=s+1}^{t} \lambda(u), \qquad \ell_s(s) = 1$$

then the nett effect is that only the formula for $P(t)$ is changed, becoming

$$P(t) = \frac{1}{\lambda(t)}[P(t-1) - \{\lambda(t) + x(t)'P(t-1)x(t)\}^{-1}P(t-1)x(t)x(t)'P(t-1)].$$

One reasonable procedure would be to take $\lambda(t) \equiv \lambda$, where λ is $0 < \lambda < 1$ and λ is fairly near to 1, e.g. 0.95.

However it is felt that h and n might be made to depend on t. In particular even if the true system were of the known order, n, then h will have to increase with t in order that $\tau(t)$ will converge to the true τ. Of course if h increases with t as $\log t$, as it will if AIC or BIC is used to choose h, then eventually the calculation cannot be done in real time. However if "forgetting" is used then the sample size is not, truly, increasing with t and thus h should not increase indefinitely. The criterion should be

$$\log \hat{\sigma}_h^2(t) + h \log f(t)/f(t) \tag{7.3}$$

where, when "forgetting" is used $f(t)$ measures the sample size to time t and is

$$f(t+1) = \lambda(t+1)f(t) + 1,$$

since the effective sample size is

$$f(t) = \sum_{s=1}^{t} \prod_{s+1}^{t} \lambda(u).$$

It remains to describe how to compute $\hat{\sigma}_h^2(t)$ in (7.3) when h is allowed so to vary. Though these procedures are described for RLS, where h indicates the order, and for $p = 1$ it will be readily seen that they can be used in the same way for AML or RML, with n taking the place of h, and for $p > 1$. Indeed they could also fairly easily be generalised to $q > 1$. Call $x_h(t)$ the vector $x(t)$ when this has been rearranged as

$$(-y(t-1),u(t-1),-y(t-2),u(t-2),\ldots,-y(t-h),u(t-h))$$

and rearrange $\tau(t)$ accordingly, calling it $\tau_h(t)$. Put

$$X_h(t)' = [x_h(1),\ldots,x_h(t)], \qquad v(t)' = (y(1),\ldots,y(t)).$$

If Q is orthogonal and

$$Q[X_h(t)v(t)] = \begin{bmatrix} R_h(t) & r_h(t) \\ 0 & s_h(t) \end{bmatrix},$$

where $R_h(t)$ is upper triangular, then $R_h(t)\tau_h(t) = -r_h(t)$ and $f(t)^{-1}s_h(t)^2$ is $\sigma_h^2(t)$. Moreover, as now will be indicated, the calculations may be done so that all $\sigma_h^2(t)$, $h = 1,\ldots,H$ may be obtained at little cost. Indeed consider

$$S = \begin{bmatrix} R_H(t) & r_H(t) \\ x_H(t+1)' & y(t+1) \end{bmatrix}$$

and construct Q as $Q_{2H}Q_{2H-1} \cdots Q_2Q_1$ where Q_i, orthogonal, acts only on rows i, $2H+1$ and introduces a zero in the $(2H+1,i)$'th place in $Q_iQ_{i-1} \cdots Q_1S$. Then if the rows numbered i, $(2h+1)$ in $Q_{i-1}Q_{i-2} \cdots Q_1S$ are

$$(0,0,\ \ldots,\ 0,\ d^{\frac{1}{2}},\ d^{\frac{1}{2}}r_2,\ \ldots\ ,\ d^{\frac{1}{2}}r_{2h+1})$$

$$(0,0,\ \ldots,\ 0,\ \delta_{i-1}^{\frac{1}{2}}x_1,\ \delta_{i-1}^{\frac{1}{2}}x_2,\ \ldots\ ,\ \delta_{i-1}^{\frac{1}{2}}x_{2h+1})$$

and d_i, x_i', δ_i are defined by the recursion

$$d_i = d + \delta_{i-1}x_i^2\ , \qquad \delta_i = d\delta_{i-1}/d_i, \qquad \delta_0 = 1$$

$$e = d/d_i, \qquad s = \delta_{i-1}x_i/d_i$$

$$x_k' = x_k - x_i r_k, \qquad r_k' = cr_k + sx_k, \qquad r_1 = r_1' = 1, \quad x_1' = 0$$

then the i, $(2h+1)$ rows of $Q_i Q_{i-1} \ldots S$ are

$$(0,0,\ \ldots,\ 0,\ d_i^{\frac{1}{2}},\ d_i^{\frac{1}{2}}r_2',\ \ldots\ ,\ d_i^{\frac{1}{2}}r_{2h+1}')$$

$$(0,0,\ \ldots,\ 0,\ \ 0,\ \delta_i^{\frac{1}{2}}x_2',\ \ldots\ ,\ \delta_i^{\frac{1}{2}}x_{2h+1}').$$

Moreover the bottom right hand element of $Q_{2H}Q_{2H-1} \ldots Q_1$ is $\dot{\delta}_{2H}^{\frac{1}{2}}$ and the bottom right hand element of $Q_{2H}Q_{2H-1} \ldots Q_1 S$ is $\delta_{2H}^{\frac{1}{2}}\hat{e}_H(t)$, where $\hat{e}_H(t)$ is $\hat{e}(t)$ for $h = H$, in RLS. Thus given

$$s_H(t)^2 = s_H(t-1)^2 + \hat{e}_H(t)^2$$

we may find $\sigma_H^2(t)$ recursively. Moreover the calculation gives $\hat{e}_h(t)^2$ for all $h \leqslant H$ at no extra cost, since the bottom right hand element of $Q_{2h}Q_{2h-1} \ldots Q_1 S$ is $\delta_{2h}^{\frac{1}{2}}\hat{e}_h(t)$ and that of $Q_{2h}Q_{2h-1} \ldots Q_1$ is $\delta_{2h}^{\frac{1}{2}}$. Thus (7.3) may be computed for all $h \leqslant H$, $\hat{h}$ chosen to minimise this, and the whole calculation made recursive. Precisely the same thing may be done with n in AML, RML. This then defines an on-line form of the algorithms in section 5, at least for $q = 1$. How useful this algorithm will be remains to be seen. If h, n are fixed it seems certainly to have some virtues. If it was believed that a rational transfer function system would fit the data well and the system was not evolving then we should set $\lambda(t) \equiv 1$ and h should be allowed to increase with t and could be chosen via (7.3). It would then, eventually increase as $\log t$. This means that eventually the algorithm could not run in real time. However h would be small compared to its long run value until it got very large and $\log t$ itself increases slowly so that the algorithm could often be run in real time up to values so large (say $t = 2000$) that it could be regarded as a real time algorithm.

Allowing h and n to increase needs investigation but could prove useful, with $\lambda(t)$ adroitly varied, to model an evolving phenomenon or even a non-linear, episodic phenomenon. When h or n varies it is likely that occasionally they will change appreciably from one value of t to another. This is because (7.3) is likely to be flat near its minimum or even have several minima near to equality. This may not matter much since all of the competing models are behaving about equally well but could be misinterpreted as evolution.

Notes on References. The field of on-line calculation is extensively surveyed in Ljung and Söderstrom (1983). The basic procedure of this section, for h, n fixed, was suggested in Mayne, Astrom and Clarke (1983) and the procedure for h, n varying in Hannan, Kavalieris and Mackisack (1986).

References.

Adamyan, V.M., Arov, D.F. and Krein, M.G. (1971) Analytic properties of Schmidt pairs for a Hankel operator and the generalised Schur-Takagi problem. Maths USSR Sbornik, 15, 31-73.

Akaike, H. (1969) Fitting autoregressive models for prediction. Ann. Inst. Stat. Math. 6, 416-431.

Akaike, H. (1969,a) Canonical correlation analysis of time series and the use of an information criterion. In: System Identification, Advances and Case Studies, eds. R.K. Mehra and D.G. Lainiotis, Academic Press, New York, 29-91.

Anderson, B.D.O. and Moore, J.B. (1979) Optimal Filtering, Prentice Hall, Englewood Cliffs.

Casti, J,L. (1977) Dynamical Systems and Their Applications, Academic Press, New York.

Cooley, J.W. and Tukey, J.W. (1965) An algorithm for machine calculation of complex Fourier Series. Mathematics of Computation, 19, 297-301.

Diestler, M. and Hannan, E.J. (1981) Some properties of the parameterization of ARMA systems with unknown order. J. Multivariate Anal. 11, 474-484.

Friedlander, B. (1982) Lattice filters for adaptive processing. Proc. IEEE, 70, 830-867.

Glover, K. (1983) All optimal Hankel norm approximations of linear multivariable systems and their L^∞ error bounds. Research Report, Control and Management Systems Division, Dept. of Engineering, Cambridge, England.

Hannan, E.J. (1970) Multiple Time Series, Wiley, New York.

Hannan, E.J. (1980) The estimation of the order of an ARMA process. Ann. Statist. 8, 1071-1081.

Hannan, E.J. (1981) Estimating the dimension of a linear system. J. Multivariate Anal. 11, 459-473.

Hannan, E.J. (1982) Testing for autocorrelation and Akaike's criterion. In: Essays in Statistical Science, eds. J.M. Gani and E.J. Hannan, Applied Probability Trust, Sheffield, 403-412.

Hannan, E.J. and Kavalieris, L. (1984) Multivariate linear time series models. Adv. Appl. Prob. 16, 492-561.

Hannan, E.J. and Kavalieris, L. (1986) Regression, autoregression models. J. Time Series Anal. 7.

Hannan, E.J., Kavalieris, L. and Mackisack, M. (1986) Recursive estimation of linear systems. Biometrika, 73, no.1.

Hannan, E.J. and Rissanen, J. (1982) Recursive estimation of mixed autoregressive moving-average order. Biometrika, 69, 81-94.

Kailath, T. (1980) Linear Systems, Prentice Hall, Englewood Cliffs.

Jewell, N.P. and Bloomfield, P. (1983) Canonical correlations of past and future for time series: definitions and theory. Ann. Statist. 11, 837-847.

Jewell, N.P. and Bloomfield, P. (1983,a) Canonical correlations of past and future for time series: bounds and computation. Ann. Statist., 11, 848-855.

Ljung, L. and Söderstrom, T. (1983) Theory and Practice of Identification, MIT Press, Cambridge, Mass.

Mayne, D.Q., Astrom, K.J. and Clarke, J.M. (1983) A new algorithm for recursive identification of parameters in controlled ARMA processes. Research Report, Dept. of Electrical Engineering, Imperial College, London.

Rissanen, J. (1983) Universal prior for parameters and estimation by minimum description length. Ann. Statist. 6, 416-431.

Shibata, R. (1980) Asymptotically efficient selection of the order of the model for estimating parameters of a linear process. Ann. Statist. 8, 147-164.

Tunnicliffe Wilson, G. (1969) Factorization of the covariance generating function of a pure moving-average process. SIAM J. Numer. Anal., 6, 1-7.

Tunnicliffe Wilson, G. (1972) The factorization of matricial spectral densities. SIAM J. Appl. Math., 23, 420-426.

Whittle, P. (1963) On the fitting of multivariate auto-regressions and the approximate canonical factorization of a spectral density matrix. Biometrika, 50, 129-134.

Chapter 2

Linear Errors-in-Variables Models

Manfred Deistler

1. Introduction

In this contribution we are concerned with some aspects of the identification problem for linear systems where both inputs and outputs are subject to ("observational") errors. Models of this kind are called errors-in-variables (EV) models.

The conventional setting in the statistical analysis of linear systems is to attribute all errors to the outputs, or (for our purposes) equivalently to add the errors to the equations. This gives the errors in equations (EE) models.

Let $\hat{x}_t$ and $\hat{y}_t$ denote the "true" inputs and outputs respectively and let x_t and y_t denote the observed inputs and outputs, then the situation can be illustrated as follows: EV models are of the form:

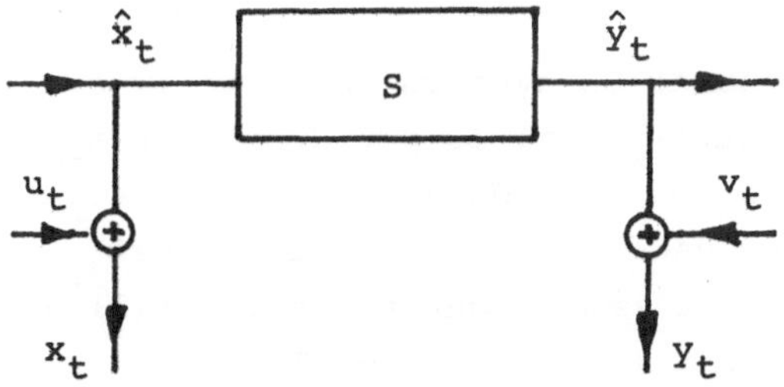

Fig 1: Schematic representation of an EV model

There u_t and v_t are the errors of the inputs and the outputs respectively. On the other hand EE models are of the form:

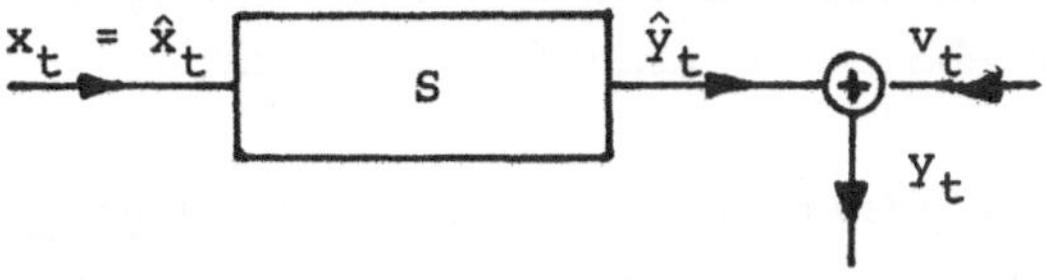

Fig 2: Schematic representation of an EE model

Of course the EV setting is more general than the EE setting. For a number of purposes, e.g. for the prediction of the observed outputs from observed inputs, the EE setting is adequate. In many cases however, the EV setting seems to be more appropriate, e.g.

(i) if our main interest concerns the "true" system generating the data (rather than a good representation of the data) and if we cannot be sure a priori that the true inputs are not contaminated by errors

(ii) if we want to decouple the common effect between the variables from the individual effects

(iii) if there is no a priori classification of the observed variables into inputs and outputs and if thus a symmetric treatment of the variables would be appropriate.

We are dealing here only with linear systems in a stationary context. Also, if the contrary has not been stated explicitely, we restrict ourselves to the single input - single output case. Our primary interest is in the characteristics of the system, i.e. in the transfer function (or the parameters of the transfer function); but also the characteristics of the errors and of $(\hat{x}_t)$ are of interest.

The statistical theory of linear dynamic EE systems, especially of ARMAX systems (also in the multi input - multi output case) has reached a certain stage of completeness now (see Hannan and Kavalieris (1984)). In the EV case on the other hand there is still a great number of open problems and this is the reason why there is still a relatively small number of applications in this field. The main problems in the EV case arise from the fact that the (ensemble) second moments of the observations do in general not uniquely determine the transfer function of the system. Another difference to EE models is, that in the EV case, higher order moments (in the non Gaussian case) may contain additional information about the transfer function.

Our emphasis is on two problems: The first is the problem of <u>identifiability</u>, i.e. the problem whether <u>the characteristics of interest</u>

mentioned above can be uniquely determined from certain characteristics of the observations as e.g. from their (ensemble) second moments or from their probability law (see Deistler and Seifert (1978)). If the answer is negative then the second problem is to describe the sets of observationally equivalent characteristics of interest, i.e. the sets of characteristics of interest which correspond to the same characteristics of the observations.

These questions are questions preceding estimation in the narrow sense and as has been stated already they turn out to be the main difficulty in the process of estimation (or inference) in EV models. This difficulty is the reason why not very much attention has been paid to EV models for a long time. However, in the last decade there has been a resurging interest in EV models in econometrics, statistics and system theory, see e.g. Aigner and Goldberger (1977), Aigner et al. (1984), Anderson B.D.O. (1985), Anderson and Deistler (1984), Anderson T.W. (1984), Deistler (1984), Deistler (1985a),Fuller (1980), Green and Anderson (1985), Hinich and Weber (1984), Kalman (1982), Kalman (1983), Maravall (1979), Picci (1985), Söderström (1980), Schneeweiß und Mittag (1985), Wegge (1983).

The paper is organized as follows. In section 2 we repeat some well known results for the static case. In sections 3 to 5 we consider the (dynamic) case when the characteristics of the observations considered are their second moments. Thereby in section 3 the set of all transfer functions corresponding to given second moments of the observations is described. Section 4 deals with the same problem, when the system is a priori known to be causal and with the problem whether causality can be detected from the second moments of the observations. In section 5 several conditions for identifiability are given. Finally in section 6 we derive conditions for identifiability using information coming from moments of order greater than two.

The system considered is of the form

$$\hat{y}_t = w(B)\hat{x}_t \qquad (1.1)$$

where B is a complex variable as well as the backward-shift operator on $\mathbb{Z}$ and where

(1.2) $$w(B) = \sum_{i=-\infty}^{\infty} w_i B^i$$

is the transfer function. The summation on the l.h.s of (1.2) ranges over all integers and thus in general the system is not a priori assumed to be causal.

The observed processes (x_t) and (y_t) are given by

(1.3) $$x_t = \hat{x}_t + u_t$$

(1.4) $$y_t = \hat{y}_t + v_t$$

We assume throughout:

(1.5) All processes considered are (wide sense) stationary; all limits of random variables are understood in the sense of mean squares convergence

(1.6) $$E\hat{x}_t = Eu_t = Ev_t = 0$$

(1.7) $$E\hat{x}_s u_t = E\hat{x}_s v_t = 0 \qquad \forall s,t$$

and

(1.8) (u_t, v_t) has a spectral density, $\tilde{f}$ say.

These assumptions are called the standard assumptions here and they will not be further explicitely restated.

The assumption $E\hat{x}_t = 0$ is imposed for notational convenience only and may easily be relaxed. (1.7) is natural in our context. Also the assumption (1.8) is natural for errors.

In many cases we in addition assume

(1.9) $Eu_s v_t = o \quad \forall s,t$

i.e. $\tilde{f}$ is diagonal

(1.10) All processes considered have a spectral density

Thereby, if (z_t) is a stationary process, we often use f_z to denote its spectral density.

Assumption (1.9) means that all common (linear) effects between (x_t) and (y_t) are due to the system and that only individual effects are attributed to the errors. Of course situations may occur where such an assumption can not be justified, e.g. if the errors in the measurement devices for inputs and outputs are correlated. Without any additional assumption the situation is hopeless because "too many" systems then correspond to given second moments of the observations. Additional information to separate the errors could be obtained from certain frequency domain properties of the errors, or from higher order moments.

2. The Static Case

Here we consider the special case, where the system is static, i.e. the transfer function w is simply the slope parameter of a line and all processes are white noise. This case has been discussed in great detail in the literature, see e.g. Gini (1921), Frisch (1934) and the surveys by Madansky (1959), Moran (1971), Aigner et al. (1984) and T.W. Anderson (1984). For the multivariable case, which is much more complicated see Kalman (1982) and Klepper and Leamer (1984).

The static EV model is written as

(2.1) $\hat{y}_t = a\hat{x}_t \quad ; a \in \mathbb{R}$

(1.3) and (1.4), where $(\hat{x}_t)$, (u_t) and (v_t) are white noise and thus

$$E\hat{x}_s\hat{x}_t = \delta_{st} \cdot \sigma_{\hat{x}} \; ; \quad Eu_s u_t = \delta_{st}\sigma_u \quad ; \quad Ev_s v_t = \delta_{st}\sigma_v$$

In addition we assume (1.9) i.e. $Eu_s v_t = o$. If we try to write

(2.1)(1.3)(1.4) as a "regression" in the observed variables, we obtain:

$$y_t = ax_t + (v_t - au_t)$$

But here $Ex_t(v_t - au_t) = -a.\sigma_u$ and thus in general (ordinary) least squares estimators will not be consistent. Therefore we have to investigate the problem in more detail.

The parameters of interest are $\theta = (a, \sigma_{\hat{x}}, \sigma_u, \sigma_v)$. The relation between these parameters and the second moments of the observations is given by

(2.2) $$\sigma_x = Ex_t^2 = \sigma_{\hat{x}} + \sigma_u$$

(2.3) $$\sigma_{xy} = Ex_t y_t = E\hat{x}_t \hat{y}_t = a.\sigma_{\hat{x}}$$

(2.4) $$\sigma_y = Ey_t^2 = a^2\sigma_{\hat{x}} + \sigma_v$$

Thus the problem of identifiability from second moments for this model is whether θ is uniquely determined from σ_x, σ_{xy}, σ_y. A slightly more general model would be of the form

(2.5) $$b\hat{y}_t = a\hat{x}_t$$ (where a and b are suitably normalized e.g. by $a^2 + b^2 = 1$)

(1.3) - (1.4). Sloppy speaking here we allow for the case $a = \infty$ in (2.1). Then the problem of observational equivalence is equivalent to the following "Frisch" problem (see Kalman (1982)): Given the covariance matrix $K = \begin{pmatrix} \sigma_x, \sigma_{xy} \\ \sigma_{yx}, \sigma_y \end{pmatrix}$ find all decompositions

(2.6) $$K = \hat{K} + \tilde{K}$$

into covariance (i.e. symmetric, nonnegative definite) matrices $\hat{K}$ and $\tilde{K}$, such that $\hat{K}$ is singular and $\tilde{K}$ is diagonal. This equivalence is straightforward; here $\hat{K}$ is the covariance matrix of

$(\hat{x}_t, \hat{y}_t)$, a and b, after suitable normalization, are defined from the linear dependence relations in $\hat{K}$, and

$$\tilde{K} = \begin{pmatrix} \sigma_u & 0 \\ 0 & \sigma_v \end{pmatrix}$$

In the case (2.1) (which excludes the possibility b = 0 in (2.5) and which is the only one we treat here, unless the contrary has been explicitely stated)

$$\hat{K} = \sigma_{\hat{x}} \cdot \begin{pmatrix} 1 & a \\ a & a^2 \end{pmatrix}$$

holds.

By the singularity of $\hat{K}$ we have

(2.7) $$\det \hat{K} = \sigma_{\hat{x}} \cdot \sigma_{\hat{y}} - \sigma_{xy}^2 = 0,$$

where $\sigma_{\hat{y}} = E\hat{y}_t^2$ and furthermore

(2.8) $$0 \leq \sigma_{\hat{x}} \leq \sigma_x$$

(2.9) $$0 \leq \sigma_{\hat{y}} \leq \sigma_y$$

and these are the only restrictions on $\sigma_{\hat{x}}$ and $\sigma_{\hat{y}}$. Thus the range of pairs $(\sigma_{\hat{x}}, \sigma_{\hat{y}})$ compatible with the given second moments of the observations is a part of a hyperbola, as illustrated in Fig 3

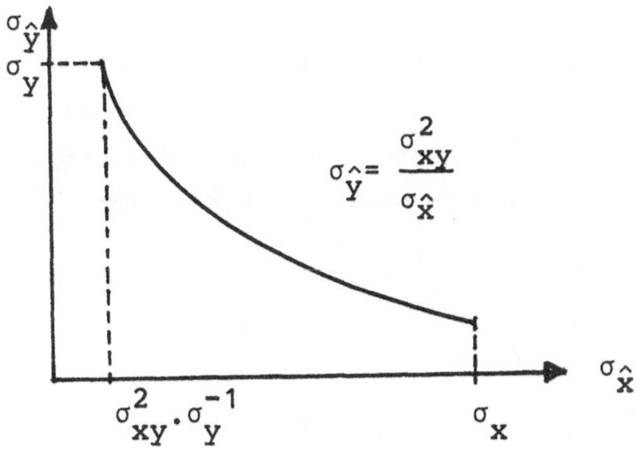

Fig 3: The range of compatible pairs $(\sigma_{\hat{x}}, \sigma_{\hat{y}})$

We do assume throughout that

(2.10) $\sigma_{\hat{x}} > 0$

and that

(2.11) $\det K > 0.$

Then the range of compatible slope parameters $a = \sigma_{xy} \cdot \sigma_{\hat{x}}^{-1}$ is given by the intervals

$$[\sigma_{xy} \cdot \sigma_x^{-1}, \ \sigma_y \cdot \sigma_{xy}^{-1}] \qquad \text{for } \sigma_{xy} > 0$$

(2.12) $$[\sigma_y \cdot \sigma_{xy}^{-1}, \ \sigma_{xy} \cdot \sigma_x^{-1}] \qquad \text{for } \sigma_{xy} < 0$$

$$\{0\} \qquad \text{for } \sigma_{xy} = 0$$

Note that the end points of these intervalls correspond to the coefficients of the (theoretical) regressions from x_t to y_t and from y_t to x_t respectively. This is an appealing result as the EV model contains the two regressions as extreme cases (where either $u_t = 0$ or $v_t = 0$).

The set of compatible parameters σ_u and σ_v is obvious from (2.2) and (2.4).

From what we said above, it also follows that <u>every</u> covariance matrix K can be decomposed as in (2.6).

Let us summarize:

<u>Theorem 2.1</u>: For every covariance matrix K there exists a corresponding EV system (2.5)(1.3)(1.4) satisfying (1.9). Under the additional assumptions (2.1), (2.10) and (2.11), the set of parameters $\theta = (a, \sigma_{\hat{x}}, \sigma_u, \sigma_v)$ compatible with given second moments of the observations is given by

(2.13) $$\{\theta = (a, a^{-1}\sigma_{xy}, \sigma_x - a^{-1}\sigma_{xy}, \sigma_y - a\sigma_{xy}) \in \mathbb{R}^4 \,|$$

$$a \in (\text{sign } \sigma_{xy}).[\,|\sigma_{xy}/\sigma_x^{-1}|, |\sigma_y./\sigma_{xy}^{-1}|\,]\} \quad \text{for } \sigma_{xy} \neq 0$$

and

$$\{\theta = \{(0, \sigma_{\hat{x}}, \sigma_x - \sigma_{\hat{x}}, 0) \in \mathbb{R}^4 \,|\, 0 < \sigma_{\hat{x}} \leq \sigma_x\} \quad \text{for } \sigma_{xy} = 0$$

This result is due to Gini (1921) and Frisch (1934).

If we drop the assumption $\sigma_{\hat{x}} \neq 0$ and consider the decomposition (2.6) (i.e. the more general case (2.5)) then we have the following picture: In the case $\sigma_{xy} \neq 0$, $\sigma_{\hat{x}} \neq 0$ must hold and in every decomposition (2.6), $\hat{K}$ must have rank equal to one. For otherwise $\hat{K} = 0$ and $K = \tilde{K}$ would not be diagonal. If $\sigma_{xy} = 0$ and $K \neq 0$, then $\hat{K}$ may either have rank equal to one, or $\hat{K} = 0$ and thus $K = \tilde{K}$. In the latter case the errors correspond to a maximal extraction of individual factors of the observed variables.

For singular K, the Frisch problem is trivial, because then $K = \hat{K}$ defines the unique decomposition (2.6), whenever $\sigma_{xy} \neq 0$.

If x_t and y_t are not necessarily one dimensional then an analogous Frisch problem can be formulated. Then the main problems are to determine the maximum corank, m^* say, of $\hat{K}$ among all decompositions (2.6) of K (i.e. to determine the maximum number of linear relations between the true variables) and to characterize the set of all (suitably normalized) linear relations corresponding to given K. These problems have not yet been solved for the general, multivariable case (see Kalman (1982), Klepper and Leamer (1984)) and they will not be treated here.

Now, let us turn again to the one dimensional case. In the case of Gaussian observations (x_t, y_t), there is no information from the data exceeding the information obtained from the second moments; thus e.g. for the slope parameter a there is an "interval of uncertainty"; therefore, of course, the model is not identifiable in this case.

There are several possibilities to overcome this "basic" nonidentifiability of the EV model. Again the reader is refered to the survey papers by Madansky (1959) and Moran (1971). As easily seen, if σ_u or σ_v or $\sigma_u . \sigma_v^{-1}$ are known then we have identifiability. Assumptions of this kind may be justified in physical applications where either the properties of the measurement instruments are a-priori known, or where the measurements can be repeated, whereas the true variables are kept constant. However for most applications such assumptions cannot be justified.

Another possibility is to utilize information coming (in the non Gaussian case) from moments of order greater than two (Geary (1942) Reiersøl (1950)). Here, for simplicity, we assume throughout that all moments up to a suitable large order exist. For technical reasons, we deal with cumulants rather than with moments (see e.g. Kendall and Stuart (1969)). If $z_1 \dots z_n$ are random variables, then their joint n-th order cumulant $c_{z_1 \dots z_n}$ is given by the coefficient of $(i)^n t_1 \dots t_n$ in the Taylor series expansion of $\ln E \exp i \sum_{j=1}^{n} z_j t_j$ about the origin. $c_{z_1 \dots z_n}$ has the following properties (see e.g. Brillinger (1981)):

$$c_{z_1} = Ez_1 \; ; \qquad c_{z_1 z_2} = E(z_1 - Ez_1)(z_2 - Ez_2)$$

$c_{z_1 \dots z_n}$ is symmetric in its arguments.

$$c_{(a_1 z_1 + a_2 z_2)\, z_3 \dots z_n} = a_1\, c_{z_1 z_3 \dots z_n} + a_2\, c_{z_2 z_3 \dots z_n}$$

if the random variables $(z_1 \dots z_n)$ and $(w_1 \dots w_n)$ are independent then

$$c_{(z_1 + w_1) \dots (z_n + w_n)} = c_{z_1 \dots z_n} + c_{w_1 \dots w_n}$$

Now, in addition we assume

(2.14) $\hat{x}_t, t \in \mathbb{Z}$ are independent and identically distributed and so are the u_t and the v_t and the processes $(\hat{x}_t)$, (u_t) and (v_t) are mutually independent.

Then we have the following relations between the n-th order cumulants:

(2.15) $c_{x^n} = c_{\hat{x}^n} + c_{u^n}$

(2.16) $c_{y^r x^{(n-r)}} = c_{\hat{y}^r \hat{x}^{(n-r)}} + c_{v^r u^{(n-r)}} = c_{\hat{y}^r \hat{x}^{(n-r)}} =$

$= a.c_{\hat{y}^{(r-1)} \hat{x}^{(n+1-r)}} = a.c_{y^{(r-1)} x^{(n+1-r)}}$; $r = 2..n-1$

(2.17) $c_{y^n} = c_{\hat{y}^n} + c_{v^n}$

where we have used the notation $c_{z^r w^{n-r}} = c_{\underbrace{z...z}_{r\ \text{times}}\underbrace{w...w}_{n-r\ \text{times}}}$ and the fact that $c_{v^r u^{n-r}} = 0$, for $r > 0$, $n-r > 0$; since v^r and u^{n-r} are independent.

Let us assume that $\hat{x}_t$ is <u>non Gaussian</u> (and that $\sigma_{\hat{x}} > 0$). Then there is a $n > 2$ such that $c_{\hat{x}^n} \neq 0$. If in addition we assume $a \neq 0$ then, using (2.16), $c_{\hat{y}^{r-1}\hat{x}^{(n+1-r)}} \neq 0$, and

(2.18) $$a = \frac{c_{y^r x^{(n-r)}}}{c_{y^{(r-1)} x^{(n+1-r)}}} \quad ; \quad r - 1 > 0, \quad n - r > 0$$

and thus a is uniquely determined from the cumulants of the observed processes. Note that for $n > 2$ (as opposed to the case $n = 2$) there are at least two cumulants of the form $c_{y^r x^{(n-r)}}$, $r > 0$, $n - r > 0$ and for these (2.16) holds. Once a is determined, $\sigma_{\hat{x}}$, σ_u and σ_v can be uniquely determined from (2.2) - (2.4).

Thus we have shown:

<u>Theorem 2.2</u>: Consider the static EV model (2.1)(1.3)(1.4) together with the assumption (2.14). Then, under the assumptions $\sigma_{\hat{x}} \neq 0$, x_t is non Gaussian and $a \neq 0$, the model is identifiable.

This theorem can be extended to the multivariate case in a straigthforward manner.

If instead of assuming that (u_t) and (v_t) are independent we postulate that $(u_t\ v_t)$ is Gaussian then $c_{u^r v^{n-r}} = 0$ whenever $n > 2$ and thus (2.16) holds for all r and for all $n > 2$ and therefore a is unique provided that there is a $n > 2$ such that $c_{\hat{x}^n} \neq 0$.

Now, let us make a few remarks on estimation: The negative logarithmic Gaussian likelihood function of the static model (2.1)(1.3)(1.4) is of the form (where constants have been neglected):

$$(2.19)\qquad L_T(\theta) = T \log \det K(\theta) + \sum_{t=1}^{T} (x_t, y_t) . K^{-1}(\theta) . (x_t, y_t)'$$

Thereby $K(\theta)$ is the covariance matrix K in (2.6) corresponding to the parameters θ, and T is the sample size. Thus the corresponding maximum likelihood estimator (MLE), obtained by minimizing $L_T(\theta)$ in (2.19) is given by

$$(2.20)\qquad K_T = \begin{pmatrix} \sigma_{x,T}, & \sigma_{xy,T} \\ \sigma_{xy,T}, & \sigma_{y,T} \end{pmatrix} = \frac{1}{T} \sum_{t=1}^{T} \binom{x_t}{y_t} (x_t, y_t)$$

The corresponding set of parameters, i.e. the estimate of the set of all observationally equivalent parameters corresponding to the true K, then according to theorem 2.1 is given by (for the case $\sigma_{xy,T} > 0$ say)

$$\{\theta = (\hat{a}, \hat{a}^{-1}.\sigma_{xy,T}, \sigma_{x,T} - \hat{a}^{-1}.\sigma_{xy,T}, \sigma_{y,T} - \hat{a}.\sigma_{xy,T})$$

$$\hat{a} \in [\sigma_{xy,T}.\sigma_{x,T}^{-1}, \sigma_{y,T}.\sigma_{xy,T}^{-1}]\}$$

In the non Gaussian case, of course (2.18) can be used to define a consistent estimator for a from the sample cumulants. Here the problem arises which cumulants should be selected and also information coming from sample cumulants of different order can be combined. The reader is referred to Drion (1951) and Scott (1950). Note also, that the estimators of obtained from (2.18) do not necessarily satisfy the restrictions coming from the second moments.

3. Second Moments and Dynamic Models: The General Case

From now on, linear dynamic systems are considered. In this and in the two sections following, only the information coming from the second moments of the observed processes (x_t, y_t) is used.

For the moment let x_t and y_t be not necessarily one dimensional. Let $z_t = (x_t, y_t)$, $\hat{z}_t = (\hat{x}_t, \hat{y}_t)$, $w_t = (u_t, v_t)$.

The general form of a linear dynamic system is:

$$(3.1) \qquad \lim_{N\to\infty} \sum_{i=-N}^{N} \bar{w}_i^{(N)} \hat{z}_{t-i} = 0, \ \bar{w}_i \in \mathbb{R}^{mxn} \ ; \ m < n$$

where

$$\bar{w}(B) = \lim_{N\to\infty} \sum_{i=-N}^{N} \bar{w}_i^{(N)} B^i \not\equiv 0$$

is the corresponding transfer function. In (3.1), analogously to (2.5) for the static case, we allow for a completely symmetric treatment of the variables. In this section we assume that the spectral densities

$$f = \begin{pmatrix} f_x & f_{xy} \\ f_{yx} & f_y \end{pmatrix}, \ \hat{f} = \begin{pmatrix} \hat{f}_x & f_{xy} \\ f_{yx} & \hat{f}_y \end{pmatrix}$$

and $\tilde{f}$ of (x_t, y_t), $(\hat{x}_t, \hat{y}_t)$ and of (u_t, v_t) respectively exist and that (1.9) holds in the sense that $\tilde{f}$ is diagonal. Then for an EV system (3.1)(1.3)(1.4) we have a decomposition analogous to (2.6):

$$(3.2) \qquad f = \hat{f} + \tilde{f}$$

where $\hat{f}$ is singular (since $\bar{w}(e^{-i\lambda})\hat{f} = 0$ by (3.1)) and $\tilde{f}$ is diagonal. Note that a matrix $f: [-\pi, \pi] \to \mathbb{C}^{nxn}$ is a spectral density of a (real) stationary process if and only if it is an integrable, nonnegative definite Hermitian matrix satisfying $f(\lambda) = f(-\lambda)'$. Conversely if we commence from f, every decomposition (3.2) where $\hat{f}$ is a singular and where $\tilde{f}$ is a diagonal spectral density matrix (we always omit the statement λ-a.e), corresponds to an EV system (3.1)(1.3)(1.4): Since $\hat{f}$ is singular, a transfer function can be found satisfying

$$(3.3) \qquad \bar{w}(e^{-i\lambda}) \ \hat{f}(\lambda) = 0$$

Thereby always a suitable normalization can be chosen such that the limit in (3.1) exist.

As easily can be seen, for every spectral density matrix f a decomposition (3.2) exists; take e.g. $\tilde{f}(\lambda) = \lambda_{min}(\lambda).I$, where $\lambda_{min}(\lambda)$ is the smallest eigenvalue of $f(\lambda)$

Clearly, for given f, in general the decomposition (3.2) is not unique. Analogously to the static multivariate case the main problem then is to determine the maximum possible number of (independent) linear dynamic relations between the $\hat{z}_t$, m*say, over all decompositions for given f and to describe the set of all observationally equivalent $\bar{w}$. Note that (3.2) may also be formulated as a dynamic factor analysis model.

Now again we restrict ourselves to the case n = 2, i.e. when x_t and y_t are one-dimensional: We assume

$$(3.4) \qquad f_{\hat{x}}(\lambda) > 0 \qquad \forall \lambda$$

and that (3.1) can be written as $\sum_{i=-\infty}^{\infty} \bar{w}_i \hat{z}_{t-i} = 0$.

Clearly, for n = 2, $\hat{f}(\lambda)$ can either have rank one or rank zero. In the second case $\hat{f}(\lambda) = 0$ and $f_{xy}(\lambda) = 0$. Imposing (3.4) implies $m^* = 1$ and $\hat{f}(\lambda)$ has rank 1 for all λ. (3.4) is automatically fulfilled if $f_{xy}(\lambda) \neq 0$ for all λ. Thus under (3.4) the system can always be written as (1.1) where $\bar{w} = (w,-1)$ is unique for given $\hat{f}$ and $f_{xy}(\lambda) = 0$ implies $w(e^{-i\lambda}) = 0$.

If f itself is singular, then $\hat{f}=f$ and $\tilde{f}=0$ defines a decomposition corresponding to an error-free system. This decomposition is unique whenever $f_{xy}(\lambda) \neq 0$. For λ's where $f_{xy}(\lambda)=0$ we may have e.g. $f_y(\lambda)=0$, ($f_x(\lambda) > 0$ and $f_{\hat{x}}(\lambda) > 0$), and $f_u(\lambda) > 0$ gives rise to another decomposition. Of course in this case we have for the corresponding transfer function $w(e^{-i\lambda})=0$. For the rest of the paper we always assume that $f(\lambda)$ is nonsingular on a set of Lebesgue measure greater than zero.

Besides the transfer function w, the other characteristics of interest are f_x, f_u, f_v.

Analogously to the static case, the set of pairs $(f_{\hat{x}}, f_{\hat{y}})$ compatible with given f satisfies

(3.5) $$0 < f_{\hat{x}} \leq f_x, \quad f_{\hat{x}}(\lambda) = f_{\hat{x}}(-\lambda); \quad f_{\hat{x}} \text{ is measurable}$$

(3.6) $$0 \leq f_{\hat{y}} \leq f_y, \quad f_{\hat{y}}(\lambda) = f_{\hat{y}}(-\lambda); \quad f_{\hat{y}} \text{ is measurable}$$

and, since $\hat{f}$ is singular

(3.7) $$|f_{xy}|^2 = f_{\hat{x}} f_{\hat{y}}$$

and (3.5)(3.6)(3.7) are the only restrictions on $(f_{\hat{x}}, f_{\hat{y}})$. Thus we have (Anderson and Deistler (1984) Deistler(1985a)).

<u>Theorem 3.1</u>: Consider the linear dynamic EV system (1.1),(1.3),(1.4). Under the additional assumptions (1.9)(1.10) and (3.4), the set of all transfer functions w satisfying

(3.8) $$|f_{yx}(\lambda)| . f_x^{-1}(\lambda) \leq | w(e^{-i\lambda})| \leq f_y(\lambda) . |f_{xy}(\lambda)|^{-1}$$
$$\varphi(w(e^{-i\lambda})) = \varphi(f_{yx}(\lambda)) \qquad \text{for } f_{xy}(\lambda) \neq 0$$

where $\varphi(z)$ denotes the phase of the complex number z, and

(3.9) $$w(e^{-i\lambda}) = 0 \qquad \text{for } f_{xy}(\lambda) = 0$$

is the set of all transfer functions w corresponding to given f

The corresponding set of the other characteristics of interest, $f_{\hat{x}}$, f_u, f_v satisfies the following relations

(3.10) $$f_{\hat{x}}(\lambda) = \begin{cases} f_{yx}(\lambda) . w^{-1}(e^{-i\lambda}) & \text{for } f_{xy}(\lambda) \neq 0 \\ 0 < f_{\hat{x}}(\lambda) \leq f_x(\lambda) & \text{for } f_{xy}(\lambda) = 0 \end{cases}$$

(3.11) $$f_u = f_x - f_{\hat{x}}$$

(3.12) $$f_v = f_y - w(e^{-i\lambda}).f_{xy}$$

Theorem 3.1 is the dynamic analogon to Theorem 2.1. It shows that the phase of the transfer function is uniquely determined from f whereas the gain of w may vary in a band whose boundaries (which depend on frequency) correspond to the dynamic regressions when either $u_t = 0$ or $v_t = 0$.

4. Causality

We here discuss two aspects of causality in the EV setting: First we consider the kind of additional identifying information obtained from an a priori causality assumption, i.e. if in (1.2) $w_i = 0, \quad i < 0$; (and thus the summation is ranging from zero to infinity only). Second, the problem of what can be said about the causality status of the system, given the second moments of the observations is treated.

Let us introduce some notation: For a polynomial

$$p(B) = \sum_{i=0}^{\delta p} p_i B^i, \qquad p_i \in \mathbb{R}, \ B \in \mathbb{C}$$

by δp we denote its degree and by p^* we denote the rational function

$$p^*(B) = p(B^{-1}) = \sum_{i=0}^{\delta p} p_i B^{-i}$$

Let δp_o denote the multiplicity of the zero of $p(B)$ at $B = 0$; by $\tilde{p}(B)$ we denote the polynomial defined as

$$\tilde{p}(B) = p^*(B).B^{\delta p}$$

$\tilde{p}$ has degree equal to $\delta p - \delta p_o$. If $B_1 \dots B_{\delta p - \delta p_o}$ are the nonzero roots of p, then $B_1^{-1} \dots B_{\delta p-\delta p_o}^{-1}$ are the roots of $\tilde{p}$. $p_o \neq 0$ implies $\tilde{\tilde{p}} = p$.

Furthermore, we define p^+ and p^- respectively by

$$p = p^+.p^-.B^{\delta p_o}$$

and

$$p^+(B) \neq 0 \ |B|<1, \ p^-(B) \neq 0, \ |B| \geq 1; \ p^-(0) = 1$$

Let $p_1, \dots p_4$ be polynomials. As well known every rational function f of the form

$$f(e^{-i\lambda}) = p_1(e^{-i\lambda}) p_2(e^{-i\lambda})^{-1} p_3^*(e^{-i\lambda}) p_4^*(e^{-i\lambda})^{-1}$$

defined on the unit circle of the complex plane has a unique rational extension to $\mathbb{C}$, given by

$$f(B) = p_1(B) p_2(B)^{-1} p_3^*(B) p_4^*(B)^{-1}$$

In this section we assume that the transfer function w and spectra considered are rational, i.e.

$$w(B) = a^{-1}(B) b(B) \tag{4.1}$$

$$f_{\hat{x}} = d^{-1} e \ \sigma_\varepsilon e^* \ d^{-1*}$$

$$f_u = c^{-1} h \ \sigma_\mu h^* \ c^{-1*} \tag{4.2}$$

$$f_v = f^{-1} g \ \sigma_\nu g^* \ f^{-1*}$$

where

$$a(B) = \sum_{i=0}^{\delta a} a_i B^i, \qquad b(B) = \sum_{i=0}^{\delta b} b_i B^i,$$

$$d(B) = \sum_{i=0}^{\delta d} d_i B^i, \qquad e(B) = \sum_{i=0}^{\delta e} e_i B^i;$$

$$c(B) = \sum_{i=0}^{\delta c} c_i B^i, \qquad h(B) = \sum_{i=0}^{\delta h} h_i B^i;$$

$$f(B) = \sum_{i=0}^{\delta f} f_i B^i, \qquad g(B) = \sum_{i=0}^{\delta g} g_i B^i$$

are polynomials, where in (4.2) we consider the spectra to be defined (by the rational extension described above) on $\not{C}$ (rather than on $|B| = 1$ or on $[-\pi,\pi]$) and where a factor of 2π has been omitted and where we in addition assume:

(4.3) $\quad a(B) \neq 0 \quad |B| = 1; \quad d(B) \neq 0 \quad |B| \leq 1; \quad e(B) \neq 0 \quad |B| \leq 1$

$\quad c(B) \neq 0 \quad |B| \leq 1; \quad h(B) \neq 0 \quad |B| < 1$

$\quad f(B) \neq 0 \quad |B| \leq 1; \quad g(B) \neq 0 \quad |B| < 1$

such that $\sigma_\varepsilon, \sigma_\mu, \sigma_\nu$ are the respective innovation variances.

$e(B) \neq 0 \; |B|=1$ then is equivalent to $f_{\hat{x}}(B) \neq 0 \; |B|=1$ i.e. to (3.4)
And finally it is assumed:

(4.4) a,b are relatively prime and so are d,e and c,h and f,g.

(4.5) $a^+(0) = d(0) = e(0) = c(0) = h(0) = f(0) = g(0) = 1$

If we impose (3.4), then the assumptions (4.3)(4.4)(4.5) are costless in the sense they do not restrict the class of transfer functions w and spectra $f_{\hat{x}}$, f_u, f_v considered, but serve only as norming conditions to obtain unique parameters. It should be stressed that concerning b we do neither impose the miniphase assumption $b(B) \neq 0 \quad |B|<1$, nor do we assume $b(0) = 1$. The assumption $a(B) \neq 0$, $|B| = 1$ guarantees the existence of a stationary solution (1.1) of $a(B)\hat{y}_t = b(B)\hat{x}_t$.

(4.6) $\quad a(B) \neq 0, \; |B| \leq 1$

then is the causality assumption.

The rationality assumptions often are imposed, even if we do not a priori know that the true transfer function and spectra are rational, in order to approximate the true quantities by quantities which can be described by a finite number of parameters.

Note that even if f is rational, the rationality assumption for the matrices $\hat{f}$ and $\tilde{f}$ in the decomposition (3.2) is an additional a-priori restriction, i.e. in general there are also non-rational $\hat{f},\tilde{f}$. This is easily seen as, e.g. for $f_x(\lambda)>0$ $f_y(\lambda)>0$, det $f>0$, $\forall\lambda$, every $f_{\hat{x}}$ such that $\frac{|f_{xy}(\lambda)|^2}{f_y(\lambda)} \leq f_{\hat{x}}(\lambda) \leq f_x(\lambda)$, where $f_{\hat{x}}$ is measurable, defines a decomposition (3.2) via (3.7), and clearly $f_{\hat{x}}$ can be chosen so to be non-rational, as, by the nonsingularity of $f(\lambda)$, the strict inequality

$$\frac{|f_{xy}(\lambda)|^2}{f_y(\lambda)} < f_x(\lambda)$$

holds. On the other hand, if f is rational, a rational decomposition $\hat{f},\tilde{f}$ always exists: Take for instance $f_{\hat{x}} = f_x$ then (3.7), (3.11) and (3.12) define rational f_y, f_u and f_v.

Consider the following example (Anderson and Deistler (1984)):

Let

$$\begin{aligned} f_x(\lambda) &= c_1 > 0 \\ f_{xy}(\lambda) &= c_2\cdot(1+be^{-i\lambda})^* \qquad |b| < 1 \quad c_1 > c_2 > 0 \\ f_y(\lambda) &= |1 + be^{-i\lambda}|^2 . c_2+c_3; \; c_3 > 0 \end{aligned} \tag{4.7}$$

Then the set of all feasible $f_{\hat{x}}$ is given by

$$\frac{c_2^2}{c_2+c_3|1+be^{-i\lambda}|^{-2}} \leq f_{\hat{x}}(\lambda) \leq c_1 , \qquad f_{\hat{x}} \text{ measurable}$$

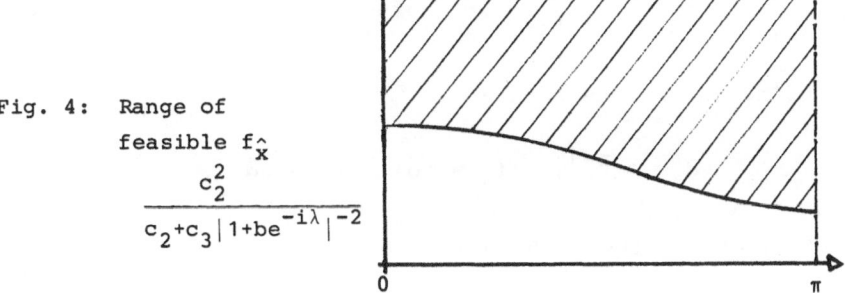

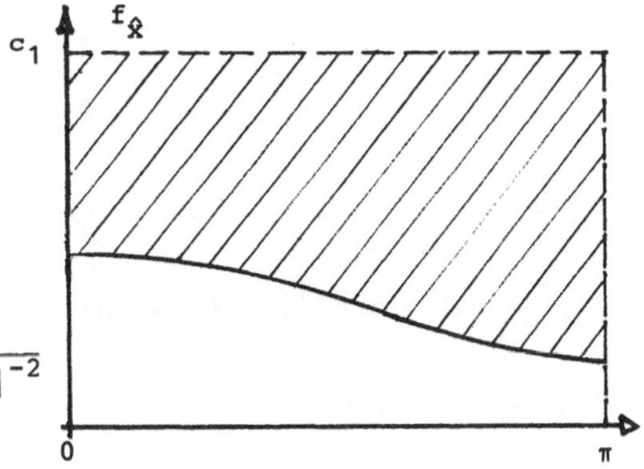

Fig. 4: Range of feasible $f_{\hat{x}}$

$$\frac{c_2^2}{c_2+c_3|1+be^{-i\lambda}|^{-2}}$$

The range of feasible $f_{\hat{x}}$ in this case containes all rational spectral densities of the form

$$f_{\hat{x}}(\lambda) = \frac{c.\prod_{1}^{p}(e^{-i\lambda}-z_j).\prod_{1}^{p}\overline{(e^{-i\lambda}-z_j)}}{c_o\prod_{1}^{q}(e^{-i\lambda}-w_j).\prod_{1}^{q}\overline{(e^{-i\lambda}-w_j)}}$$

for arbitrarily chosen $|z_j|<1$, $|w_j|<1$, p and q and for suitably chosen c, c_o. Without restriction of generality we assume $z_j \neq w_i$.

Under the rationality assumptions of this section, the relation between the second moments of the observations and the characteristics of interest is of the form:

(4.8) $$f_x = d^{-1}e\ \sigma_\varepsilon e^*d^{-1*} + c^{-1}h\ \sigma_\mu h^* c^{-1*}.$$

(4.9) $$f_{xy} = a^{-1}b\ d^{-1}e\ \sigma_\varepsilon e^* d^{-1*}$$

(4.10) $$f_y = a^{-1}b\ d^{-1}e\ \sigma_\varepsilon e^* d^{-1*} b^* a^{-1*} + f^{-1}g\ \sigma_\nu g^* f^{-1*}$$

In a first step, we analyse the information about w (and $f_{\hat{x}}$) coming from f_{yx}: If $\bar{w}$, $\bar{f}_{\hat{x}}$ corresponds to an EV system also satisfying (4.9) then we have:

(4.11) $$\bar{w} = w.f_{\hat{x}}.\bar{f}_{\hat{x}}^{-1} = c f_2 \tilde{f}_2 f_1^{-1} \tilde{f}_1^{-1} B^{\delta f_1 - \delta f_2}\ w$$

(4.12) $$\bar{f}_{\hat{x}} = c^{-1} f_1 \tilde{f}_1 f_2^{-1} \tilde{f}_2^{-1}.B^{\delta f_2 - \delta f_1} f_{\hat{x}}$$

where, using an obvious notation

$$c = \sigma_\varepsilon \bar{\sigma}_\varepsilon^{-1}, \quad f_1 = \bar{e}d, \quad f_2 = e\bar{d}$$

and thus f_i are polynomials satisfying

(4.13) $$f_i(B) \neq 0 \qquad |B| \leq 1 \ ; \quad i = 1,2$$

(4.14) $\qquad f_i(0) = 1 \qquad ; \qquad i = 1,2$

and

(4.15) $\qquad c > 0.$

Conversely, if (4.13) - (4.15) hold, then (4.11) and (4.12) obviously defines a $\bar{w}$ and $\bar{f}_{\hat{x}}$ giving f_{yx} via (4.9) (and satisfying our assumptions)

Now, let us write $w = b^+b^- B^{\delta b_o}(a^+a^- B^{\delta a_o})^{-1}$.
Then $n = \delta b^- + \delta b_o - \delta a^- - \delta a_o$ is the number of zeros of w of modulus less than one minus the number of poles of w of modulus less than one. From (4.11) and (4.13) we see that n is uniquely determined for given f_{yx}. Thus we have shown (i) and (ii) in the following theorem:

<u>Theorem 4.1</u>: Let the assumptions (4.1) - (4.5) hold. Then:

(i) w and $\bar{w}$ are transfer functions corresponding to the same f_{yx} if and only if there exist polynomials f_1, f_2 satisfying (4.13) (4.14) and a constant $c > 0$ such that (4.11) holds

(ii) $n = \delta b^- + \delta b_o - \delta a^- - \delta a_o$ is an invariant for all EV systems with the same f

(iii) If $n = 0$, then there is a transfer function w (corresponding to f_{xy}) which is both causal and miniphase

(iv) If $n > 0$, then there is a transfer function which is causal, but no transfer function which is miniphase.

(v) If $n < 0$, then there is no causal transfer function, but there is a transfer function which is miniphase

(vi) If it is a priori known that the transfer functions are causal, i.e. if (4.6) holds, then $w = a^{-1}b$ and $\bar{w}$ correspond to the same f_{yx} if and only if there exists a polynomial f_2 satisfying (4.13) (4.14) and

(4.16) $\quad 0 \leq \delta f_2 \leq n$

and a constant $c > 0$ such that

(4.17) $\quad \bar{w} = c.(b^+\tilde{a}^-)(a^+\tilde{b}^-)^{-1}.f_2\tilde{f}_2.B^{n-\delta f_2}$

holds

Proof: If in (4.11) we take $f_2 = \tilde{a}^-$ and $f_1 = \tilde{b}^-$ then

$$\bar{w} = c(b^+\tilde{a}^-)\;(a^+\tilde{b}^-)^{-1}B^n$$

where $(b^+\tilde{a}^-(a^+\tilde{b}^-)$ is a causal and miniphase transfer function; together with (ii) this implies (iii) - (v) and also (vi) is easily seen.

This result has been stated in Deistler (1985a) and Deistler (1985 b). Partly more general results have been given in Anderson, B.D.O. (1985) for the causal, non necessarily rational single input - single output case and in Green and Anderson (1985) for a causal multivariable case. From (4.17) we see that the causality assumption gives a substantial reduction of the set of all transfer functions compatible with given f_{yx}. If $n = 0$, then in the causal case, w is unique up to multiplication by a positive constant (this has been pointed out by Hinich (1983) and Anderson, B.D.O. (1985)). An estimation procedure for the case $n=0$ has been developed by Hinich (1983) and Hinich and Weber (1984).

5. Conditions for Identifiability from the Second Moments of the Observations

This section consists of two parts: In the first part, we investigate the additional information coming from (4.8) and (4.10). So this is the second step of the analysis of the previous section for the rational case. Special emphasis is put on identifiability. In the second part we give some other conditions for identifiability.

We have (see Anderson and Deistler (1984), Anderson, B.D.O. (1985), Deistler (1985a)(1985b), Maravall (1979), Nowak (1983)):

Theorem 5.1: Let the assumptions (4.1) - (4.5) hold. Then:

(i) If the transfer functions are a priori known to be causal and if either $n = 0$ or if

(5.1) $d, \tilde{b}^-$ are relatively prime

(5.2) a, e are relatively prime

(5.3) $b^+, \tilde{b}^-$ are relatively prime

then w is uniquely determined from f under each of the following conditions:

(5.4) d, c are relatively prime and $\delta d > 0$

(5.5) $a.d, f$ are relatively prime and $\delta ad > 0$

(5.6) $\delta d = 0$ and $\delta e > \delta h - \delta c$

(5.7) $\delta ad = 0$ and $\delta e + \delta b > \delta g - \delta f$

(ii) If w is a priori assumed to be causal and if d and c are a priori assumed to be relatively prime, then there is only a finite number of factors f_2 in (4.14) compatible with given f_x.

(iii) If the transfer functions are not necessarily causal, and if the assumptions (5.1) - (5.3)

(5.8) $a, \tilde{e}$ are relatively prime

(5.9) $a^+, \tilde{a}^-$ are relatively prime

hold, then under (5.4) or (5.5) or (5.6) or (5.7) w is unique

(iv) If $w, f_{\hat{x}}$ corresponds to given f_{yx}, then all cw, $c^{-1} f_{\hat{x}}$, where c satisfies

(5.10) $$0 < c_{min} \leq c \leq c_{max}$$

with c_{min} and c_{max} defined by

(5.11) $$\min_{|B|=1} \; (f_x(B) - c_{min}^{-1} f_{\hat{x}}(B)) = 0$$

and

(5.12) $$\min_{|B|=1} \; (f_y(B) - c_{max} |w(B)|^2 f_{\hat{x}}(B)) = 0$$

correspond to given f, and for all other c, cw, $c^{-1}f_x$ does not correspond to given f.

Proof: (i): If n = 0, then as has already been stated, w is uniquely determined from (4.9) up to multiplication by a positive constant. The same holds under (5.1) - (5.3): Due to (5.1)and(5.2) no pole - zero cancellations on the r.h.s in (4.9) can occur and thus a and d are uniquely determined from the poles in f_{yx}. By (5.3), e then is uniquely determined from those zeros of add*f_{yx},B_i say, where also B_i^{-1} is a zero and thus b and w are unique up to multiplication by a positive constant.

From (4.8) we have

$$(5.13) \qquad df_x d^* = e\sigma_\varepsilon e^* + dc^{-1}h\sigma_\mu h^* c^{-1*} d^*$$

If (5.4) holds, then there exists at least one zero of d, B_1 say, and we have

$$df_x d^*(B_1) = e\sigma_\varepsilon e^*(B_1)$$

and from this σ_ε and thus b are uniquely determined. The proof for (5.5) is completely analogous.

If (5.6) holds then (4.8) is of the form

$$f_x = e\sigma_\varepsilon e^* + c^{-1}h\sigma_\mu h^* c^{-1*}$$

and thus c is uniquely obtained from the poles of f_x. Then σ_ε is obtained from a comparison of coefficients of power $\delta e + \delta c$ in

$$cf_x c^* = ce\sigma_\varepsilon e^* c^* + h\sigma_\mu h^*$$

and in the same way we proceed if (5.5) holds.

The proof of (iii) is completely analogous, since (5.1) - (5.3)(5.8) (5.9) here again guarantee that w is determined from f_{yx} up to multiplication by a positive constant.

(ii) if d and c are relatively prime, then all zeros of d are poles of f_x and thus there is only a finite number of candidates for d and thus also for f_2 in (4.17)
(iv) is an immediate consequence of Theorem 4.1 and of (4.8) and (4.10), taking into account the non-negativity of spectral densities for $|B| = 1$

Clearly, once w is uniquely determined and if $w(e^{-i\lambda}) \neq 0$ then also $f_{\hat{x}}$, f_u, f_v are unique. (i) and (iii) show that e.g. using (4.19), once the degrees are prescribed and if $\delta d > 0$, we have identifiability on a generic subset of the parameter space.

Now we discuss some other cases where additional a priori restrictions guarantee identifiability from the second moments of the observations

(i) Let the inputs $\hat{x}_t$ have a spectral distribution function $F_{\hat{x}}$ given by

$$(5.14) \qquad F_{\hat{x}}(\lambda) = \int_{[-\pi,\lambda]} f_{\hat{x}} d\lambda + \sum_{j:\lambda_j \leq \lambda} F_{x,j} \quad ; \quad F_{x,j} > 0$$

Thus (x_t) is a fairly general process where F_x has an absolutely continuous and a discrete part and where the discrete part corresponds to a stationary harmonic process $\Sigma e^{i\lambda_j t} z_{x,j}$, where $F_{x,j} = E|z_{x,j}|^2$. Here we do not impose (1.9). By assumption (1.8), (u_t,v_t) has a spectral density and thus we have

$$(5.15) \qquad F_x(\lambda) = \int_{[-\pi,\lambda]} (f_{\hat{x}}+f_u) d\lambda + \sum_{j:\lambda_j \leq \lambda} F_{x,j}$$

and

$$(5.16) \qquad F_{yx}(\lambda) = \int_{[-\pi,\lambda]} w(e^{-i\lambda}) dF_{\hat{x}}(\lambda) + \int_{[-\pi,\lambda]} f_{vu} d\lambda$$

$$= \int_{[-\pi,\lambda]} w(e^{-i\lambda})(f_{\hat{x}}+f_u) d\lambda + \sum_{j:\lambda_j \leq \lambda} w(e^{-i\lambda_j}) F_{x,j} + \int_{[-\pi,\lambda]} f_{uv} d\lambda$$

Thus, from the jumps $F_{x,j}$ and $F_{xy,j}$ in F_x and F_{xy} we obtain:

(5.17) $$w(e^{-i\lambda_j}) = F_{xy,j} \cdot F_{x,j}^{-1}$$

If $w = a^{-1}.b$ is rational with prescribed (maximal) degrees na and nb say for a and b respectively then w is determined under our assumptions from na+nb+1 (different) values $(\lambda_j, w\,(e^{-i\lambda_j}))$ $j = 1 \ldots na+nb+1$. Clearly this result can be extended to the multivariate case.

(ii) If it is a priori known, that the input-errors have the property

$$f_u(\lambda) = 0 \qquad \forall \lambda \in A \subset [0,\pi]$$

where A is a nonempty open set then clearly $f_{uv}(\lambda) = 0$ and $f_{\hat{x}}(\lambda) = f_x(\lambda)$ $\forall \lambda \in A$, and therefore, provided that $f_{\hat{x}}(\lambda) \neq 0$, $\lambda \in A$, we have

(5.18) $$w(e^{-i\lambda}) = f_{yx}(\lambda) f_{\hat{x}}(\lambda)^{-1} \qquad \lambda \in A$$

If w is rational, then it is uniquely determined from (5.18), since A is open, e.g. by the derivatives of $w(e^{-i\lambda})$ at some point $\lambda \in A$.

(iii) From (4.8) we see that if c=1, d is uniquely determined from f_x: In this case the poles of f_x which are located outside the unit circle are the zeros of d and as d(0)=1, d is unique.

If (1.9) holds, if the input errors u_t are white noise (i.e. $c^{-1}.h=1$) and if d e, then w is unique (Söderström (1980)). From (4.8) we obtain:

(5.19) $$d\, f_{\hat{x}} d^* = e_{\vartheta} \sigma_{\varepsilon} e^* + d \sigma_{\mu} d^*$$

As d is uniquely determined from f_x, a comparison of coefficients corresponding to power δd in (5.19) gives $\sigma\mu$. Then the usual factorization of $(f_{\hat{x}} - \sigma_{\mu})$ into factors that have no poles inside or on and no zeros inside the unit circle gives $d^{-1}e$ and σ_{ε} and thus also $f_{\hat{x}}$ and w.

(iv) If (1.9) holds, if $(\hat{x}_t)$ is autoregressive (i.e. e=1) and (u_t) is a moving average process (i.e. c=1) and if $\delta d > 0$ (i.e. (x_t)) is autoregressive in the narrow sense, not white noise) then w is unique (Söderström (1980)). Here (4.8) is of the form:

$$f_{\hat{x}} = d^{-1}\sigma_\varepsilon d^{-1*} + h\sigma_\mu h^*$$

Again, d is uniquely determined from f_x. Now

$$d f_{\hat{x}} d^* = \sigma_\varepsilon + dh\sigma_\mu h^* d^*$$

and if B_1 is a zero of d, we obtain

$$\sigma_\varepsilon = d f_{\hat{x}} d^*(B_1)$$

and thus $f_{\hat{x}}$ and w are unique.

A common feature of the cases (i) - (iv) is that $f_{\hat{x}}$ is obtained from (4.8) and the extra assumptions on the spectra $f_{\hat{x}}$ and f_u imposed; once $f_{\hat{x}}$ is known, the uniqueness of w, f_u, f_v is immediate. Analogous conditions may be imposed to detect $f_{\hat{y}}$ from (4.10). Once $f_{\hat{y}}$ is unique, we have

$$w = f_{\hat{y}} \cdot f_{xy}^{-1}$$

provided that $f_{xy}(\lambda) \neq 0$, $\forall\lambda$ and the rest is easy.

6. Identifiability from High Order Moments

For non Gaussian observations, analogously to the static case, information coming from moments of order greater than two may be useful for identifiability (Akaike (1966), Deistler (1986)). In addition to our standard assumptions we here assume:

(6.1) $(\hat{x}_t)$ and (u_t, v_t) are strictly stationary processes; the processes are (mutually) independent and all moments up to order n, where n is sufficiently large, exist and the cumulants of $(\hat{x}_t)$ and of $(\hat{y}_t)$ satisfy conditions of the form

$$\sum_{t_1..t_{n-1}=-\infty}^{\infty} |c_{\hat{y}_{t_1}\hat{y}_{t_2}..\hat{y}_{t_r}\hat{x}_{t_{r+1}}...\hat{x}_{t_{n-1}}x_o}| < \infty$$

and the same holds for (u_t) and (v_t)

Then (see e.g. Brillinger (1981)) the corresponding n-th order cumulant spectrum exists and is given by

(6.2) $$f_{\hat{y}^r\hat{x}(n-r)}(\lambda_1...\lambda_{n-1}) =$$

$$= (2\pi)^{-n+1} \cdot \sum_{t_1..t_{n-1}=-\infty}^{\infty} c_{\hat{y}_{t_1}..\hat{y}_{t_r}\hat{x}_{t_{r+1}}..\hat{x}_{t_{n-1}}\hat{x}_o} \exp\{-i \sum_{j=1}^{n-1} \lambda_j t_j\}$$

and analogously for (u_t) and (v_t). As easily seen, due to linearity and continuity of cumulants with respect to one variable (when the others are kept constant), we obtain from (6.2) and (1.1):

(6.3) $$f_{\hat{y}^r\hat{x}(n-r)}(\lambda_1...\lambda_{n-1}) =$$

$$= (2\pi)^{-n+1} \cdot \sum_{t_1..t_{n-1}=-\infty}^{\infty} \left(\sum_{i=-\infty}^{\infty} w_i c_{\hat{x}_{t_1-i}\hat{y}_{t_2}..\hat{y}_{t_r}..\hat{x}_o}\right) \exp\{-i \sum_{j=1}^{n-1} \lambda_j t_j\} =$$

$$= w(e^{-i\lambda_1}) \cdot f_{y(r-1)x(n-r+1)}(\lambda_1...\lambda_{n-1})$$

Furthermore, from the properties of the cumulants we obtain

(6.4) $$f_{y^r x^{n-r}}(\lambda_1...\lambda_{n-1}) = f_{\hat{y}^r\hat{x}(n-r)}(\lambda_1...\lambda_{n-1}) +$$

$$+ f_{v^r u(n-r)}(\lambda_1...\lambda_{n-r})$$

If, in addition we assume

(6.5) (u_t) and (v_t) are independent

then

(6.6) $f_{v^r v(n-r)}(\lambda_1...\lambda_{n-r}) = 0 \quad$ for $r > 0, \quad n - r > 0$

If we assume

(6.7) $(u_t\ v_t)$ is Gaussian,

then $f_{v^r u}(n-r) = 0$ for all $n > 2$. Thus we have

<u>Theorem 6.1</u>: Consider the (dynamic) EV-model (1.1)(1.3)(1.4). If in addition the assumptions (6.1), (6.5) and

(6.8) $$f_{y(r-1)x(n-r+1)}(\lambda_1\lambda_2\dots\lambda_{n-1}) \neq 0 \qquad \forall\lambda_1,$$

for suitable $\lambda_2\dots\lambda_{n-1}$ and suitable $n>2$; $r-1>0$, $n-r>0$ are satisfied

then w is uniquely determined from

(6.9) $$w(e^{-i\lambda_1}) = f_{y^r x(n-r)}(\lambda_1\dots\lambda_{n-1}) \cdot f^{-1}_{y(r-1)x(n-r+1)}(\lambda_1\dots\lambda_{n-1})$$

An analogous result holds if (6.5) is replaced by (6.7).

Theorem 6.1 is the dynamic analogon to Theorem 2.2. If $\hat{x}_t$ has a Wold decomposition (see e.g. Hannan (1970))

$$\hat{x}_t = w_2(z)\varepsilon_t$$

where (ε_t) is i.i.d., then (see e.g. Brillinger (1981))

(6.10) $$f_{\hat{x}}(\lambda_1\dots\lambda_{n-1}) = (2\pi)^{-n+1} . w_2(e^{-i\lambda_1})..w_2(e^{-i\lambda_{n-1}}) . \\ . w_2(\exp i \sum_{i=1}^{n-1} \lambda_j) . c_{\varepsilon^n}$$

If all moments of ε_t exist if ε_t is non Gaussian, (provided that $\varepsilon_t \neq 0$) there is a $n>2$ such that $c_{\varepsilon^n} \neq 0$. If in addition $w_2(e^{-i\lambda}) \neq 0$ $\forall\lambda$, then $f_{\hat{x}^n}(\lambda_1\dots\lambda_{n-1}) \neq 0$ $\forall\lambda_1\dots\lambda_{n-1}$ and then due to (6.3) condition (6.8) is fulfilled.

The generalization of Theorem 6.1 to the multivariable case is straightforward. Estimators of the transfer function w may be obtained from (6.9) replacing the cumulant spectra by their estimators.

References

Aigner,D.J. and A.S.Goldberger (Eds.), (1977): Latent Variables in Socio-Economic Models. North Holland P.C., Amsterdam

Aigner,D.J., C.Hsiao, A.Kapteyn and T.Wansbeek (1984): Latent Variable Models in Econometrics. In: Griliches, Z. and M.D.Intriligator (Eds.) Handbook of Econometrics. North Holland P.C., Amsterdam

Akaike,H. (1966): On the Use of Non-Gaussian Process in the Identification of a Linear Dynamic System. Annals of the Institute of Statistical Mathematics 18, 269 - 276

Anderson,B.D.O. (1985): Identification of scalar errors-in-variables models with dynamics, Forthcoming in Automatica

Anderson,B.D.O. and M.Deistler (1984): Identifiability in Dynamic Errors-in-Variables models, Journal of Time Series Analysis, 5, 1-13

Anderson,T.W. (1984): Estimating Linear Statistical Relationships. Annals of Statistics, 12, 1 - 45

Brillinger,D.R. (1981): Time Series: Data Analysis and Theory. Expanded Edition. Holden Day, San Francisco

Deistler,M. (1984): Linear errors-in-variables models. In: J.Franke, W.Härdle und D.Martin (Eds.), Robust and Nonlinear Time Series Analysis, Lecture Notes in Statistics, Springer-Verlag, Berlin

Deistler,M. (1985a): Linear dynamic errors-in-variables models in: J.Gani and M.Priestley (Eds.) Essays in Time Series and Allied Processes. Forthcoming

Deistler,M. (1985b): Identifiability and Causality in Linear Dynamic Errors-in-Variables Systems. In: Proc. 5th Franco Belgian Meeting of Statisticians. Forthcoming

Deistler,M. and H.G.Seifert (1978): Identifiability and Consistent Estimability in Dynamic Econometric Models. Econometrica, 46, 969 - 980

Drion,E.F. (1951): Estimation of the Parameters of a Straight Line and of the Variances of the Variables, if they are Both Subject to Error. Indegationes Math. 13, 256 - 260

Frisch,R. (1934): Statistical Confluence Analysis by Means of Complete Regression Systems. Publication No. 5, University of Oslo, Economic Institute

Fuller,W.A. (1980): Properties of some Estimators for the Errors-in-Variables Model. Annals of Statistics, 8, 407 - 422

Geary,R.C. (1942): Inherent Relations between Random Variables. Proceedings of the Royal Irish Academy, Sec. A, 47, 63 - 76

Geary,R.C. (1943): Relations between Statistics: The General and the Sampling Problem When the Samples are Large. Proceedings of the Royal Irish Academy. Sec. A, 49, 177 - 196

Gini,C. (1921): Sull'interpolazione di una retta quando i valori della variable indipendente sono affetti da errori accidentali. Metron 1, 63 - 82

Green,M. and B.D.O.Anderson (1985): Identification of multivariable errors-in-variables models with dynamics. Mimeo.

Hannan,E.J. (1970): Multiple Time Series. Wiley, New York

Hannan,E.J. and L.Kavalieris (1984): Multivariate Linear Time Series Models. Advances in Applied Probability 16, 492 - 561

Hinich,M.J. (1983): Estimating the Gain of a Linear Filter from Noisy Data. In: D.R.Brillinger and P.R.Krishnaiah (Eds.) Handbook of Statistics, Vol 3. North Holland, Amsterdam

Hinich,M.J. and W.E.Weber (1984): Estimating Linear Filters with Errors in Variables Using the Hilbert Transform. Federal Reserve Bank of Minneapolis, Res.Dept. Staff Report 96

Kalman,R.E. (1982): System Identification from Noisy Data. In: A.Bednarek and L.Cesari (Eds.) Dynamical Systems II, a University of Florida International Symposium. Academic Press, New York

Kalman,R.E. (1983): Identifiability and Modeling in Econometrics. In: Krishnaiah,P.R. (Ed.) Developments in Statistics, Vol 4. Academic Press, New York

Kendall,M.G. and A.Stuart (1969): The Advanced Theory of Statistics. Vol 1, 3rd Edition, Griffin, London

Klepper,S. and E.Leamer (1984) Consistent Sets of Estimates for Regressions with Errors in all Variables. Econometrica 52, 163 - 183

Madansky,A. (1959): The Fitting of Straight Lines when Both Variables are Subject to Error. Journal of the American Statistical Association 54, 173 - 205

Maravall, A. (1979): Identification in Dynamic Shock-Error Models. Springer Verlag, Berlin.

Moran, P.A.P. (1971): Estimating Structural and Functional Relation ships. Journal of Multivariable Analisys 1, 232-255

Nowak, E. (1983): Identification of the Dynamic Shock-Error Model with Autocorrelated Errors. Journal of Econometrics 23, 211-221

Picci, G. (1985): Factor Analylis Models via Stochastic Realization Methods. This Volume

Reiersøl,O. (1941): Confluence Analysis by Means of Lag Moments and other Methods of Confluence Analysis. Econometrica 9, 1 - 24

Reiersøl,O. (1950): Identifiability of a Linear Relation Between Variables which are subject to Error. Econometrica 18, 375 - 389

Schneeweiß,H. und H.J.Mittag (1985): Lineare Modelle mit fehlerbehafteten Daten. Physica Verlag, Würzburg

Scott,E.L. (1950): Note on Consistent Estimates of the Linear Structural Relation Between two Variables. Annals of Mathematical Statistics 21, 284 - 288

Söderström,T. (1980): Spectral Decomposition with Application to Identification. In: Archetti,F. and M.Cugiani (Eds.) Numerical Techniques for Stochastic Systems. North Holland P.C., Amsterdam

Wegge,L. (1983): ARMAX-Models Parameter Identification without and with Latent Variables. Working Paper. Dept. of Economics, Univ. of California, Davis.

Chapter 3

A New Class of Dynamic Models For Stationary Time Series

Giorgio Picci and Stefano Pinzoni

1. Introduction

In this note we shall discuss a new class of dynamic models which may be better suited than conventional ARMAX schemes to describe non-causally interacting time series. Typical areas of application that we have in mind include econometrics (where it is often not clear what variables are "endogenous" and what are "exogenous") and identification of industrial processes operating under feedback. In these situations there is no a priori clear causality relation among the variables and, in fact, a possible goal of the identification experiment could be the testing for existence of causal relations.

The class of models introduced here is a natural dynamic generalization of the well-known static Factor Analysis model which in various equivalent forms (the most popular of which seems to be the so-called Errors-In-Variables scheme) has been object of much study in the past especially by econometricians and psychologists. (For definitions of these concepts and a rather comprehensive survey of the literature one may consult the recent paper by Van Schuppen(1985). The study of these models has recently been revitalized by Kalman in a series of papers(Kalman, 1982a, 1982b and 1983)and some of the critiques presented in Kalman's work have been the motivating stimulus for the earlier paper (Finesso and Picci, 1984). The present exposition represents the natural continuation and generalization of the results presented there. In order to improve readability we have chosen to skip some non essential technical details. A more complete story can be found in (Picci and Pinzoni, 1986). People interested in general philosophical discussions on the modelling problem considered here are referred to the introduction of (Finesso and Picci, 1984).

We should mention that some of the specific issues dealt with in this paper are also treated (in the scalar E.I.V. context) in the work of Anderson and Deistler (1984), Anderson (1985), Deistler (1985). Although the primary motivations (and hence the basic assumptions) in these papers are of a rather different nature than ours, the reader might find some ground for comparisons in the discussion of the causality problem presented in section 4.

For the sake of motivating the introduction of Dynamic Factor Analysis models we shall briefly review the definition of <u>causality</u> of a dynamical model, first in the deterministic and then in the stochastic (Gaussian) case. The idea that we want to convey is that causal models are quite "nongeneric" mathematical descriptions to impose aprioristically to real data, e.g. economic time series or data coming from industrial processes involving feedback.

In a deterministic framework the notion of causality is of course well known. Assume that the components of the m-dimensional variable y(t), whose temporal evolution is described by a certain dynamical model, have been grouped in two subvectors,

$$y(t) = \begin{bmatrix} y_1(t) \\ y_2(t) \end{bmatrix}, \tag{1.1}$$

with $y_i(t) \in \mathbb{R}^{m_i}$, $i = 1,2$, and $m_1 + m_2 = m$. It is intuitively clear that a dynamical model should quantify the <u>dynamic relation</u>

occurring between the variables y_1 and y_2 (i.e. how much y_1 "influences" y_2 and vice versa). This is made precise in J.C.Willems refoundation of Systems Theory (Willems,1979): any model (or equivalently dynamical system) with external variables y is just a subset of trajectories $\mathcal{B}$ (called the behaviour of the system) in $(\mathbb{R}^m)^{\mathbb{Z}} = (\mathbb{R}^{m_1})^{\mathbb{Z}} \times (\mathbb{R}^{m_2})^{\mathbb{Z}}$ and therefore a bona fide relation between y_1 (ranging over $(\mathbb{R}^{m_1})^{\mathbb{Z}}$) and y_2 (ranging on $(\mathbb{R}^{m_2})^{\mathbb{Z}}$). We say that y_1 causes y_2 or, equivalently, that y_1 is the input and y_2 is the output variable of the system, if this relation specializes to a very particular kind of function, namely if

$$y_2(t) = f(y_1), \quad t \in \mathbb{Z}, \tag{1.2}$$

where f depends only on the values taken by y_1 before and at time t.

In the stochastic case the sharply defined subset $\mathcal{B}$ is replaced by a probability measure on the sample space $(\mathbb{R}^m)^{\mathbb{Z}}$ and thus the external variable y becomes a stochastic process $\{y(t)\}$. The model is in this case just the probability law of $\{y(t)\}$. To make things simple we shall consider here about the simplest possible class of random processes, described in the following

BASIC ASSUMPTION

The process $\{y(t)\}$ is an m-dimensional Gaussian stationary process with zero mean and has a rational spectral density S strictly positive definite on the unit circle (i.e. $S(e^{i\theta}) > 0$). □

We shall write the spectrum S in a partitioned form corresponding to the subdivision (1.1) of the external variables,

$$S = \begin{bmatrix} S_1 & S_{12} \\ S_{21} & S_2 \end{bmatrix}, \qquad (1.3)$$

where the blocks S_i, of dimension $m_i \times m_i$, represent the auto spectra and S_{12} the cross spectrum of the two components $\{y_1(t)\}$ and $\{y_2(t)\}$ of dimension m_1 and m_2.

The definition of causality in this context, essentially due to Granger (1963 and 1969), sounds as follows.

DEFINITION 1.1

We say that the process y_1 causes y_2 or, equivalently, that y_1 is an input process with corresponding output y_2 if, for all $t \in \mathbb{Z}$,

$$E[y_2(t) \mid y_1] = E[y_2(t) \mid y_1(s),\ s \leq t], \qquad (1.4)$$

where the first conditional expectation is with respect to the whole history $\{y_1(t);\ t \in \mathbb{Z}\}$ of the component y_1.

□

Causality is just conditional independence of the past and present output history $\{y_2(s);\ s \leq t\}$ from future inputs $\{y_1(s);\ s > t\}$ given the past of the input $\{y_1(s);\ s \leq t\}$ and can of course be defined in a much more general setting than the one adopted here. In a Gaussian setting we can however translate everything in the convenient Hilbert space language of the linear theory of random processes (see e.g. Rozanov, 1967). Some of this material necessary for future use will be quickly reviewed in the next paragraphs.

We shall denote the vector space of all finite linear combinations of the scalar random variables $\{\alpha' y(t);\ \alpha \in \mathbb{R}^m,\ t \in \mathbb{Z}\}$ closed in the metric induced by the scalar product $\langle x, z \rangle := E\, x\, z$, by the symbol $H(y)$ (sometimes abbreviated to H). $H_t^-(y)$, $H_t^+(y)$ will denote the past and future subspaces spanned by the random variables y(s) up to and, respectively, after and at, time t. Clearly,

$$H_t^{\pm}(y) = U^t H_o^{\pm}(y) \tag{1.5}$$

where $U : y(t) \rightarrow y(t+1)$ is the (unitary) shift operator of the process $\{y(t)\}$. Normally the subscript zero in (1.5) will be dropped. For the two components y_1 and y_2 we shall define the subspaces $H(y_1)$, $H(y_2)$ (abbreviated to H_1 and H_2 when there is no danger of confusion) accordingly. Obviously $H = H_1 \vee H_2$ where the wedge denotes closed vector sum.

Subspaces like H_1 and H_2 are doubly invariant for the shift U, in the sense that they satisfy $U^t H_i = H_i$ for all $t \in \mathbb{Z}$. The multiplicity of a doubly invariant subspace $X \subset H$ is the cardinality of any minimal generating set, i.e. is the smallest n for which one can find random variables $\{x_1, \dots, x_n\}$ in X such that the vector space generated by $\{U^t x_i ;\ i = 1, \dots, n,\ t \in \mathbb{Z}\}$ is dense in X. The process $\{x(t)\}$ with $x_i(t) = U^t x_i$ is called a generating process of X.

By the Spectral Representation Theorem (Rozanov, 1967), there is a unitary representation of the random variables in H(x) as n-dimensional (row) vector functions in the Hilbert space $L_n^2(C, dQ)$ where $C = \{z;\ |z| = 1\}$ is the unit circle in the complex plane and Q is the $n \times n$ matrix spectral distribution measure of the process $\{x(t)\}$. Each random variable $\xi(t) := U^t \xi$ with $\xi \varepsilon X$ can be written as

$$\xi(t) = \int_{-\pi}^{\pi} e^{i\theta t} f(e^{i\theta}) d\hat{x}(e^{i\theta})$$

for a unique $f \in L_n^2(C, dQ)$. Here $\hat{x}$ is the n-dimensional random spectral measure of the stationary process $\{x(t)\}$. As it is well known (Rozanov, 1967) the spectral distribution matrix is related to $\hat{x}$ by $dQ = E(d\hat{x}\, d\hat{x}^*)$, where the star means conjugate transpose.

The representation will be symbolically written as

$$\xi(t) = f(z)x(t) . \tag{1.6}$$

The System Theoretic interpretation of the notation is that the stationary process $\{\xi(t)\}$ is obtained by passing the stationary process $\{x(t)\}$ through the linear (stable) filter of transfer function f.

In all cases of interest for us the spectral distribution measure of $\{x(t)\}$ will be absolutely continuous with respect to Lebesgue measure on C. The spectral density matrix will still be denoted by the symbol Q. It is well known (compare e.g. Fuhrmann, 1981, p. 111) that $\{x_1,...,x_n\}$ being a <u>minimal generating set</u> is equivalent to $Q(e^{i\theta})$ being <u>strictly positive definite</u> on a set of positive Lebesgue measure. For example the assumption $S(e^{i\theta}) > 0$ a.e. guarantees that $H = H(y)$ has precisely multiplicity m, a possible minimal set of generators being given by the m scalar components of the random vector y(0). Observe further that any other minimal generating process for H(x) can be written as

$$u(t) = T(z)x(t) , \tag{1.7}$$

with T an nxn matrix function having rows in $L^2_n(C,dQ)$ and Q-<u>a.e. nonsingular on the unit circle</u>.

Of course when $\{x(t)\}$ admits a spectral density Q which is a.e. positive definite on the unit circle, then all admissible T 's will be a.e. nonsingular on C and all minimal generating processes for H(x) will have an a.e. positive definite spectral density on the unit circle. In particular, by choosing $T=W^{-1}/\sqrt{2\pi}$

where W is any square solution of the standard spectral factorization problem $WW^* = Q$, we obtain white noise generators $\{u(t)\}$ for $H(x)$. The transfer function in the representation (1.6) in this case belongs to $L^2_n(C, d\theta/2\pi)$. In this context we shall call causal any function f with vanishing positive Fourier coefficients in $L^2_n(C, d\theta/2\pi)$, i.e. such that

$$\int_{-\pi}^{\pi} e^{-i\theta k} f(e^{i\theta}) d\theta/2\pi = 0 \tag{1.8}$$

for all $k>0$. Thus any causal function belongs to the n-dimensional conjugate Hardy space $\bar{H}^2_n$ (Hoffmann, 1962) and can be extended to a function of the complex variable z analytic on $\{|z| > 1\}$ (including the point at infinity). A matrix valued function T will be called causal if its rows are. It can be verified directly that for any generating process $\{x(t)\}$ with a strictly positive definite spectral density matrix(o) we have $H^-_t(u) \subset H^-_t(x)$ if and only if the transfer matrix T in (1.7) is causal. A (left -) invertible matrix T with rows in $L^2_n(C, d\theta/2\pi)$ will be called minimum phase if it is causal and its extension has an analytic (left -) inverse on $\{|z| > 1\}$. This is the same thing as a conjugate outer matrix function in H^2-theory.

We finally recall the concept of conditional orthogonality. Two subspaces H_1, H_2 of H will be said conditionally orthogonal, given a third subspace X (notation: $H_1 \perp H_2 \mid H$), if

$$< h_1 - E^X h_1, h_2 - E^X h_2 > = 0 \tag{1.9}$$

for all $h_1 \in H_1$ and $h_2 \in H_2$. Here the symbol E^X denotes orthogonal projection onto X. Since in the Gaussian case conditional expec-

(o) or, more generally, full rank purely non deterministic (Rozanov, 1967).

tation given a certain family of random variables in H is the same thing as orthogonal projection onto the subspace of H spanned by them, we see that conditional orthogonality is the same property as conditional independence, given X, of the two families H_1 and H_2 of Gaussian random variables. The concept of conditional orthogonality will be extensively used in this paper. For additional information one may consult (Lindquist and Picci, 1985).

We return to our discussion of causality in the stochastic setting. The following is a rather well known fact although often stated in a different terminology.

THEOREM 1.1

The process $\{y_1(t)\}$ causes $\{y_2(t)\}$ if and only if

$$y_2(t) = A(z)y_1(t) + v(t) , \tag{1.10}$$

where $A(z)$ is an $m_2 \times m_1$ causal matrix function and $\{v(t)\}$ a stationary process completely independent of $\{y_1(t)\}$, i.e.

$$E\, y_1(t)v'(s) = 0 \tag{1.11}$$

for all $t,s \in \mathbf{Z}$.

This result is essentially due to (Caines and Chan, 1976). It is also discussed in (Caines and Chan, 1975) and (Gevers and Anderson, 1982). In these references causality is called "absence of feedback" (from y_2 to y_1). Note that (1.10) is nothing else but the popular ARMAX scheme widely used in time series identification. Just express $\{v(t)\}$ by its innovation representation,

$$v(t) = G(z)e(t) , \tag{1.12}$$

where $G(z)$ is minimum phase, normalized so as to make $G(\infty) = I$, and $\{e(t)\}$ is a white noise process. Recall that, by rationality of $S(z)$, both $A(z)$ and the spectrum of $\{v(t)\}$ are rational and then express the rational matrix $[A(z)\ G(z)]$ by a left coprime M.F.D. $D(z)^{-1}[B(z)\ C(z)]$ to get

$$D(z)y_2(t) = B(z)y_1(t) + C(z)e(t). \tag{1.13}$$

The orthogonality condition (1.11) holds if and only if

$$E\, e(t)y_1'(s) = 0\ , \quad t,s \in \mathbf{Z}\ , \tag{1.14}$$

and therefore using ARMAX models with independent (or uncorrelated) noise and input (y_1) processes(o) is equivalent to <u>imposing a priori a causality relation on the data</u>. In this case the statistical inference problem of estimating the joint law of $\{y_1(t)\}$ and $\{y_2(t)\}$ is reduced to the much simpler problem of estimating just the conditional law of future y_2's given past inputs y_1.

Quite often there is no evidence in the data which justifies the use of causal models. What kind of models should then be used in this situation? One obvious answer would be to describe the whole (joint) process $\{y(t)\}$ by an m-dimensional ARMA scheme corresponding say to the $m \times m$ rational minimum phase spectral factor of the joint spectrum S. Our main concern is however in describing how two given groups of variables (y_1 and y_2) interact dynamically. In practice y_1 and y_2 have a precise physical or economic meaning and the main reason for doing modelling and identification is to discover how much of the temporal evolution of each variable is "explained" by the other. For this purpose it would be much more useful to have models which (although necessarily equivalent to the joint ARMA scheme mentioned above) put into explicit evidence the mutual influence of the variables y_1 and y_2. A class of mathe-

(o) Actually condition (1.14) is often considered to be part of the definition of an ARMAX model and is not even explicitly mentioned.

matical descriptions which in a certain sense generalizes the causal model (1.10) is the stochastic feedback scheme

$$\begin{aligned} y_2(t) &= L(z)y_1(t) + v_1(t), \\ y_1(t) &= K(z)y_2(t) + v_2(t), \end{aligned} \tag{1.15}$$

where L and K are causal transfer functions and $\{v_1(t)\}$ and $\{v_2(t)\}$ stationary "error" processes whose innovations can at most be assumed orthogonal to the past histories of y_1 and y_2, respectively. This class of models has been extensively investigated in recent years, especially by Gevers and Anderson (1981 and 1982) and Anderson and Gevers (1982) with the main motivation of understanding identifiability of control systems operating under feedback. Practical use of these models for time series identification seems however to have been very limited so far.

We shall propose here a different class of models in which the dynamic interaction between y_1 and y_2 is explicitly described by the introduction of an auxiliary variable x. This auxiliary variable will play a role similar to the state variable in Systems Theory.

DEFINITION 1.2

A Dynamic Factor Analysis Model with external variables the (jointly stationary) vector processes $\{y_1(t)\}$ and $\{y_2(t)\}$, is a linear relation of the form

$$\begin{aligned} y_1(t) &= A_1(z)x(t) + w_1(t), \\ y_2(t) &= A_2(z)x(t) + w_2(t), \end{aligned} \tag{1.16}$$

where $A_1(z)$ and $A_2(z)$ are transfer matrices of dimension $m_1 \times n$ and $m_2 \times n$ and $\{x(t)\}$, $\{w_1(t)\}$, $\{w_2(t)\}$ are zero mean stationary processes of dimensions n, m_1, m_2 which are pairwise uncorrelated, i.e.

$$\{w_1(t)\} \perp \{x(t)\} \perp \{w_2(t)\}. \tag{1.17}$$

□

Note that A_1 and A_2 need not be causal. The process $\{x(t)\}$ will sometimes be referred to as the factor process of the model. A Dynamic Factor Analysis (F.A.) model will be called rational if A_1, A_2 are rational matrices and $\{x(t)\}$ has rational spectrum. The terminology (although not terribly elegant) has been extrapolated from the static case.

In the next sections we shall present a first rudimentary analysis of the model (1.16). The main questions one would like to answer concern the representability of an arbitrary joint stationary process $\{y(t)\}$ (with $y(t)$ partitioned as in (1.1)) by models of the type (1.16), the equivalence of representations (i.e. when do different representations describe the same spectrum S or the same process $\{y(t)\}$), the "external behaviour" of the model which is obtained once $\{x(t)\}$ is eliminated, finding a natural notion of minimality and characterizations of minimal models, parametrizations and canonical forms in the rational case and above all discuss use of Factor Analysis models in Statistical Inference (i.e. identification). This is quite a large program and only a few of these aspects will be touched upon in this paper. Others (especially the last two mentioned above), which still need more research, will not be discussed here.

2. Dynamic Factor Analysis Models

The stationary processes $\{x(t)\}$, $\{w_1(t)\}$, $\{w_2(t)\}$ which define a Factor Analysis model span a certain Hilbert space $H(x,w_1,w_2)$ which we denote by H_o. The Factor Space X of the model (1.16) is the doubly invariant subspace of H_o generated by the factor process,

$$X = \overline{\text{span}}\ \{\alpha' x(t);\ \alpha \in \mathbb{R}^n,\ t \in \mathbb{Z}\}\ . \tag{2.1}$$

Let $\hat{n} \leq n$ be the multiplicity of X and let $\{\hat{x}(t)\}$ be a minimal generating process for X. Clearly, since $x(t) = T(z)\hat{x}(t)$ for some $n \times \hat{n}$ matrix T, we can always rewrite the model (1.16) with $A_1(z)$ and $A_2(z)$ replaced by $\hat{A}_1(z) = A_1(z)T(z)$ and $\hat{A}_2(z) = A_2(z)T(z)$ and a factor process $\hat{x}(t)$ which is a minimal generating process for X. We shall therefore adhere from now on to the convention of considering only F.A. models in which $\{x(t)\}$ is a minimal generating process for X. Hence the multiplicity of X will always coincide with the dimension of $x(t)$. Two F.A. models which differ by a change of (minimal) generators in X will be called equivalent. Obviously two equivalent models have the same $\{w_i(t)\}$ processes (for i=1,2), the same factor space X and transfer matrices and factor processes related by

$$\begin{aligned} \hat{A}_i(z) &= A_i(z)T(z)^{-1}, \qquad i = 1,2\ , \\ \hat{x}\,(t) &= T(z)x(t), \end{aligned} \tag{2.2}$$

where T is a Q-a.e. nonsingular $n \times n$ matrix function whose rows belong to $L^2_n(C,dQ)$, Q being the spectral distribution measure of $\{x(t)\}$. It is easy to check that (2.2) defines an equivalence relation on the class of all F.A. models of $\{y_1(t)\}$, $\{y_2(t)\}$.

We shall now introduce the concept of splitting subspace. By this idea we shall be able to attach a precise probabilistic meaning to F.A. models and at the same time reduce this notion to a very simple geometric object. Let $H_i = H(y_i)$, $i = 1,2$ be the Hilbert spaces spanned by the components $\{y_i(t)\}$, $i = 1,2$. It will be useful to think of H_1 and H_2 as (doubly invariant) subspaces embedded in a large Hilbert space H_o obtained by suitably augmenting $H_1 \vee H_2$. On H_o there is defined a unitary shift operator U which reduces to the shift of the process $\{y(t)\}$ on the subspace $H = H(y) = H_1 \vee H_2$. (The role played by H_o is very similar to that of the space $H(x,w_1,w_2)$ introduced at the beginning of this section).

DEFINITION 2.1

A (stationary) Splitting Subspace is a doubly invariant subspace X of H_o which makes $H(y_1)$ and $H(y_2)$ conditionally orthogonal given X, i.e. satisfies

$$H(y_1) \perp H(y_2) \mid X \tag{2.3}$$

together with $UX = X$.

A Splitting Subspace X is called minimal if there are no proper subspaces of X which are doubly invariant and still satisfy condition (2.3).

□

The concept of splitting subspace is a generalization of the idea of sufficient statistic (at least in the Gaussian case). It follows in fact from the definition of conditional orthogonality (1.9) that

$$E[h_1 \mid X \vee H_2] = E[h_1 \mid X] , \qquad h_1 \in H_1 ,$$

and, equivalently,

$$E[h_2 \mid X \vee H_1] = E[h_2 \mid X] , \qquad h_2 \in H_2 ,$$

so that all what is relevant in $H_2(H_1)$ at the purpose of predicting any $h_1 \in H_1$ $(h_2 \in H_2)$ is already contained in X. Therefore if X (or any system of generators of X) is given, we can disregard H_2 (H_1) completely. Note that the concept of splitting is of interest only if it corresponds to effective data reduction. Hence the notion of minimality is of central importance.

LEMMA 2.1 (Ruckebusch, 1976 and Lindquist and Picci, 1985)

A splitting subspace X is minimal if and only if

$$\bar{E}^X H_1 = X , \quad \bar{E}^X H_2 = X \tag{2.4}$$

(here $\bar{E}^X H_i$ is the closure of $\{E^X h_i ;\ h_i \in H_i\}$, $i = 1,2$). □

The following theorem shows that (modulo choice of generators) splitting subspaces and Dynamical Factor Analysis models are essentially the same thing.

THEOREM 2.1

The factor space X of any F.A. model of $\{y_1(t)\}$, $\{y_2(t)\}$ is a splitting subspace. Vice versa to every splitting subspace X for $H(y_1)$, $H(y_2)$ of finite multiplicity there corresponds the equivalence class, defined modulo choice of generators, of F.A. models having X as factor space.

Proof:

Let X be given by (2.1). Then, since

$$A_i(z)x(t) = E^X y_i(t) , \quad t \in \mathbf{Z},\ i = 1,2 , \tag{2.5}$$

the orthogonality relation of $\{w_1(t)\}$ and $\{w_2(t)\}$, which holds by assumption for any model (1.16), can be rewritten as

$$y_1(t) - E^X y_1(t) \perp y_2(s) - E^X y_2(s) , \qquad t,s \in \mathbf{Z} . \tag{2.6}$$

As $\{y_i(t)\}$ is a generating process for H_i it follows from the definition (1.9) that indeed X is splitting. Viceversa, let X be a splitting subspace and $\{x(t)\}$ a minimal generating process for X of dimension n. The projections $E^X y_i(t)$ can be written as in (2.5) for suitable transfer functions $A_i(z)$ of dimension $m_i \times n$. Define

$$w_i(t) := y_i(t) - E^X y_i(t), \quad t \in \mathbb{Z},\ i = 1,2, \tag{2.7}$$

then the stationary processes $\{w_i(t)\}$ are orthogonal to X and, by the conditional orthogonality condition (2.3), we have also $E\, w_1(t) w_2(s)' = 0$ for all $t,s \in \mathbb{Z}$. Therefore $\{y_1(t)\}$ and $\{y_2(t)\}$ can be written as in (1.16), while satisfying (1.17).

□

The equivalence established by Theorem 2.1 permits to define a first rough notion of minimality for F.A. models. We shall say that a F.A. model is irreducible if its factor space is minimal splitting.

THEOREM 2.2 (Picci and Pinzoni, 1986)

A F.A. model is irreducible if and only if the rank a.e. on the unit circle of the matrices $A_1(z)$ and $A_2(z)$ is equal to the multiplicity of X.

All irreducible F.A. models have the same multiplicity (i.e. the same number of factors) n equal to the rank a.e. on the unit circle of the cross spectrum S_{12} of the processes $\{y_1(t)\}$ and $\{y_2(t)\}$.

In the rest of this paper we shall concentrate on irreducible models. As we have just seen these models are characterized by a.e. left invertible matrices $A_k(z)$, $k = 1,2$. Their factor process has an absolutely continuous spectrum with an a.e. positive definite spectral density matrix Q on the unit circle (Picci and Pinzoni, 1986).

If in an irreducible F.A. model we eliminate the auxiliary variable $\{x(t)\}$, we obtain a scheme of the following type,

$$A_2(z)^{-L}\hat{y}_2(t) = A_1(z)^{-L}\hat{y}_1(t) \ , \tag{2.8}$$

$$y_1(t) = \hat{y}_1(t) + w_1(t) \ , \tag{2.9a}$$

$$y_2(t) = \hat{y}_2(t) + w_2(t) \ . \tag{2.9b}$$

This is essentially what is commonly called an Errors-In-Variables (E.I.V.) model of the processes $\{y_1(t)\}$, $\{y_2(t)\}$. Here $y_1(t)$ and $y_2(t)$ are represented as "noisy" observations of the "true" variables $\hat{y}_1(t)$, $\hat{y}_2(t)$ obeying the deterministic relation (2.8) . Note that the correlation structure of $\{y_1(t)\}$ and $\{y_2(t)\}$ is completely embodied in the relation (2.8) as the noise processes $\{w_k(t)\}$ are mutually uncorrelated and also orthogonal to the "true" variables $\{\hat{y}_k(t)\}$. An equivalent form of the deterministic link (2.8) between the true variables is obtained by substituting $x(t) = A_1(z)^{-L}\hat{y}_1(t)$ into the second equation in (1.16), getting

$$\hat{y}_2(t) = W(z)\hat{y}_1(t) \ , \quad W(z) := A_2(z)A_1(z)^{-L} \tag{2.10}$$

or, dually,

$$\hat{y}_1(t) = W(z)^{\#}\hat{y}_2(t) \ , \quad W(z)^{\#} = A_1(z)A_2(z)^{-L} \ . \tag{2.11}$$

Note that the transfer functions $W(z)$, $W(z)^{\#}$ and also the relation (2.8) are invariant under change of generators, $\hat{x}(t) = T(z)x(t)$ (T nonsingular), and are therefore uniquely attached to the (minimal) splitting subspace X of the model. An important question concerns the existence of models for which W (or $W^{\#}$) is a causal transfer function. This is the same as asking if two stationary processes described by an arbitrary joint spectrum S can be represented by the "noisy" input-output model

$$y_1(t) = \hat{y}_1(t) + w_1(t),$$
$$y_2(t) = W(z)\hat{y}_1(t) + w_2(t), \qquad (2.12)$$

where W(z) is causal and $\{w_1(t)\} \perp \{\hat{y}_1(t)\} \perp \{w_2(t)\}$. We shall take up this kind of questions in section 4.

As a last general comment about F.A. models, we remark that the freedom of changing generators in the factor space X permits to choose transfer matrices $A_k(z)$ or factor processes of very special structure. For example we can always take $\{x(t)\}$ to be a white noise process or require that both $A_1(z)$ and $A_2(z)$ be <u>causal</u> transfer functions. For simplicity we shall state the next result for the case of rational F.A. models.

PROPOSITION 2.1

<u>For every rational irreducible F.A. model there is a choice of (minimal) generators in</u> X,

$$\hat{x}(t) = T(z)x(t) ,$$

<u>which (maintains rationality and) achieves causality of the transfer function matrices</u>

$$\hat{A}_k(z) = A_k(z)T(z)^{-1}, \qquad k = 1,2 \quad . \qquad (2.13)$$

<u>Proof</u>:

In a rational model both the spectrum Q and the matrices A_k, $k = 1,2$ are rational functions. Since the joint spectrum $\hat{S}$ of the processes $\hat{y}_k(t) = A_k(z)x(t)$, $k = 1,2$,

$$\hat{S} = \begin{bmatrix} \hat{S}_1 & S_{12} \\ S_{21} & \hat{S}_2 \end{bmatrix} = \begin{bmatrix} A_1 \\ A_2 \end{bmatrix} Q \begin{bmatrix} A_1^* & A_2^* \end{bmatrix} , \qquad (2.14)$$

is then itself a rational function, it admits causal (in particular minimum phase) rational spectral factors. Note that irreducibility implies that rank $\hat{S} = n = \text{rank } S_{12}$. Pick a causal <u>full rank</u> rational spectral factor $\hat{A}$ (of dimension $m \times n$) of $\hat{S}$ and write it as a partitioned matrix with two blocks $\hat{A}_k(z)$ of dimensions $m_k \times n$, $k = 1,2$. The spectral factorization

$$\hat{S}(z) = \begin{bmatrix} \hat{A}_1(z) \\ \hat{A}_2(z) \end{bmatrix} \begin{bmatrix} \hat{A}_1(z)^* & \hat{A}_2(z)^* \end{bmatrix}$$

is clearly equivalent to the representations $\hat{y}_k(t) = \hat{A}_k(z)\hat{x}(t)$, $k = 1,2$ with $\{\hat{x}(t)\}$ an n-dimensional <u>white noise</u> process. We interpret $\{\hat{x}(t)\}$ as the new factor process of the model. Since $\hat{A}(z)$ is full rank, we can solve for $\hat{x}(t)$ in the representation

$$\begin{bmatrix} \hat{y}_1(t) \\ \hat{y}_2(t) \end{bmatrix} = \begin{bmatrix} A_1(z) \\ A_2(z) \end{bmatrix} x(t) = \begin{bmatrix} \hat{A}_1(z) \\ \hat{A}_2(z) \end{bmatrix} \hat{x}(t) \; ,$$

getting

$$\hat{x}(t) = \begin{bmatrix} \hat{A}_1(z) \\ \hat{A}_2(z) \end{bmatrix}^{-L} \begin{bmatrix} A_1(z) \\ A_2(z) \end{bmatrix} x(t) := T(z)x(t) \; .$$

Note that T is square $n \times n$ and nonsingular because of irreducibility. This proves Proposition 2.1.

□

In the proof we could in particular have chosen $\hat{A}(z) = [\hat{A}_1(z)' \; \hat{A}_2(z)']'$ minimum phase. We see that an irreducible <u>rational</u> F.A. model can always be written as a pair of ARMAX equations,

$$D_1(z)y_1(t) = B_1(z)\hat{x}(t) + C_1(z)e_1(t),$$
$$D_2(z)y_2(t) = B_2(z)\hat{x}(t) + C_2(z)e_2(t), \qquad (2.15)$$

with $\{\hat{x}(t)\}$, $\{e_1(t)\}$, $\{e_2(t)\}$ pairwise uncorrelated white noise processes and $D_k(z)$ and $C_k(z)$ stable polynomial matrices of dimensions $m_k \times m_k$ and $m_k \times p_k$, p_k being the multiplicity of the noise process $\{w_k(t)\}$, $k = 1,2$.

3. Stochastic realization

The main problem of this section will be to describe the class of all irreducible F.A. models which match a given spectral density matrix. We shall see that this is equivalent to solving the following problem.

PROBLEM P.1

Given an $m \times m$ spectral density matrix S partitioned as in (1.3) and satisfying the Basic Assumption of Sect. 1, find all 5-tuples of matrix functions $\{A_1, A_2, Q, R_1, R_2\}$ on the unit circle, with A_k of dimension $m_k \times n$ and of rank n, Q of dimension $n \times n$ and nonsingular, R_k of dimension $m_k \times m_k$, $k = 1,2$, which

i) satisfy the system of equations

$$S_1 = A_1 Q A_1^* + R_1,$$
$$S_{12} = A_1 Q A_2^*, \qquad (3.1)$$
$$S_2 = A_2 Q A_2^* + R_2,$$

ii) make the $(m+n) \times (m+n)$ matrix

$$\tilde{S} = \begin{bmatrix} S_1 & S_{12} & A_1 Q \\ S_{12}^* & S_2 & A_2 Q \\ QA_1^* & QA_2^* & Q \end{bmatrix} \tag{3.2}$$

<u>into a spectral density matrix (in particular Hermitian and non-negative definite on the unit circle)</u>.

□

Assume we have an irreducible F.A. model,

$$\begin{aligned} z_1(t) &= A_1(z)x(t) + w_1(t) \ , \\ z_2(t) &= A_2(z)x(t) + w_2(t) \ ; \end{aligned} \tag{3.3}$$

if we interpret Q as the spectral density matrix of $\{x(t)\}$ and R_k, $k = 1,2$ as the spectra of the two noise processes $\{w_k(t)\}$, we see that eqns. (3.1) express precisely the fact that the joint spectrum of $\{z_1(t)\}$ and $\{z_2(t)\}$ coincides with the given joint spectrum S. Note also that the matrix $\tilde{S}$ in (3.2) is just the joint spectral density of the three processes $\{z_1(t)\}$, $\{z_2(t)\}$ and $\{x(t)\}$. Vice versa, assume we are given a 5-tuple $\{A_1, A_2, Q, R_1, R_2\}$ of matrices satisfying eqns. (3.1) and condition (ii). It is not hard to see and we shall check this later, that condition (ii) implies that Q, R_1, R_2 are necessarily bounded Hermitian positive semidefinite (Q is actually positive definite) matrices on the unit circle and can therefore be interpreted as spectral densities of three mutually uncorrelated zero mean Gaussian processes $\{x(t)\}$, $\{w_1(t)\}$, $\{w_2(t)\}$. Starting from these processes, we generate $\{z_1(t)\}$ and $\{z_2(t)\}$ by the linear transformation (3.3). We see from (3.1) that the joint spectrum of the stationary processes $\{z_1(t)\}$, $\{z_2(t)\}$ is precisely equal to the given joint spectral density matrix S. In short, <u>solving</u>

problem P.1 is the same thing as finding all irreducible F.A. models (3.3) for which the joint spectrum of the external variables $\{z_1(t)\}$, $\{z_2(t)\}$ is equal to the given spectral density matrix S.

This problem is a distributional or "weak sense" stochastic realization problem (Finesso and Picci, 1984 and Lindquist and Picci, 1985). Interpreting S as the joint spectrum of two given Gaussian processes $\{y_1(t)\}$, $\{y_2(t)\}$, we are looking for all irreducible models (3.3) such that $\{z_k(t)\}$ and $\{y_k(t)\}$ are equal processes in distribution. In "practical" terms this means that the model (3.3) will only be useful to simulate the signals $\{y_k(t)\}$ in an "average" sense but not samplewise in general. A 5-tuple $\{A_1,A_2,Q,R_1,R_2\}$ satisfying conditions (i) and (ii) above, or, equivalently a F.A. model of the type (3.3) matching the given spectrum S, will be called a F.A. representation of the spectrum S.

A (strong sense) F.A. representation of the processes $\{y_1(t)\}$, $\{y_2(t)\}$ is instead a F.A. model of the type (3.3) for which $z_k(t) = y_k(t)$ almost surely for all $t \in Z$. This type of (samplewise) equality is clearly stronger than equality in distribution and can only occur when the processes $\{z_k(t)\}$ and $\{y_k(t)\}$ are defined on the same probability space. This means that the various processes $\{x(t)\}$, $\{w_1(t)\}$, $\{w_2(t)\}$ in (3.3) must be built in such a way that $H_o := H(x,w_1,w_2) \supset H(y_1,y_2) = H(z_1,z_2)$. Samplewise (i.e. strong sense) F.A. representations of $\{y_1(t)\},\{y_2(t)\}$ can be classified according to "how big" an underlying space H_o is needed to support the processes which specify the model. Later we shall study in some detail the class of F.A. representations for which $H_o = H(y_1,y_2)$. These representations will be called "y-measurable"(o).

(o) Clearly an equivalent condition for y-measurability is that the factor space X is included in H(y).

Note that whenever $\{x(t)\}$ is given, the noise processes $\{w_k(t)\}$ are automatically fixed as functions of $\{x(t)\}$, $\{y_1(t)\}$, $\{y_2(t)\}$ by the orthogonality condition (1.17), as

$$w_k(t) = y_k(t) - E^X y_k(t) , \qquad k = 1,2 , \tag{3.4}$$

where X is the splitting subspace generated by $\{x(t)\}$. Therefore a (strong) F.A. representation is completely specified once the factor process $\{x(t)\}$ is assigned as a function of some available generators of the space H_o. In particular a y-measurable representation is completely specified once $\{x(t)\}$ is given as a function of $\{y_1(t)\}$ and $\{y_2(t)\}$.

In order to avoid complicated statements about equivalence classes, it will be useful to fix once and for all a rule for choosing generators in each factor space X. A convenient way to do this is to fix a full rank factorization of the cross spectrum S_{12},

$$S_{12}(z) = H(z)G^*(z) , \tag{3.5}$$

where H and G are of respective dimensions $m_1 \times n$, $m_2 \times n$ and of rank equal to $n = \text{rank } S_{12}$ a.e. on C. Since S_{12} is rational, we can always choose H and G to be rational matrices. In fact we shall choose H and G in such a way that (3.5) is a minimal factorization of the rational matrix S_{12} (in the sense of Bart, Gohberg and Kaashoek (1979), p. 84).

Since all entries of a rational spectral density matrix must be analytic on the unit circle, it follows that both H(z) and G(z) must also be analytic on the unit circle. In the following we shall make the simplifying assumption that $S_{12}(z)$ has no zeros on the unit circle, i.e.

$$\text{rank } S_{12}(e^{i\theta}) = n, \qquad \forall\, \theta \in [0,2\pi) . \tag{3.6}$$

This guarantees that neither H(z) nor G(z) can have zeros

on the unit circle, more precisely, both $H(e^{i\theta})$ and $G(e^{i\theta})$ will be of constant rank n <u>for all</u> $\theta \in [0,2\pi)$. From now on the matrices H and G will be considered as data of our problem.

LEMMA 3.1

<u>Let condition</u> (3.6) <u>hold. Then for each equivalence class of irreducible F.A. models of</u> $\{y_1(t)\}$, $\{y_2(t)\}$ <u>there is a unique choice of generating process</u> $\{x(t)\}$ <u>in the factor space</u> X <u>such that</u>

$$A_1(z) = H(z), \qquad A_2(z) = G(z)Q(z)^{-1}, \tag{3.7}$$

<u>where</u> Q <u>is the (nonsingular) spectrum of</u> $\{x(t)\}$. <u>Alternatively, a unique generating process</u> $\{\bar{x}(t)\}$ <u>can be chosen for which</u>

$$A_1(z) = H(z)\bar{Q}(z)^{-1}, \qquad A_2(z) = G(z), \tag{3.8}$$

<u>where</u> $\bar{Q}$ <u>is the spectrum of</u> $\{\bar{x}(t)\}$. <u>The generating processes</u> $\{x(t)\}$, $\{\bar{x}(t)\}$ <u>for the same minimal splitting subspace</u> X <u>are related by the transformation</u>

$$\bar{x}(t) = Q^{-1}(z)x(t) . \tag{3.9}$$

<u>Proof</u>:

In fact, if we start with an arbitrary irreducible model (1.16), there is a unique change of generators in X, $\hat{x}(t) = T(z)x(t)$, with T such that $H(z)T(z) = A_1(z)$. Note that there is a unique a.e. nonsingular solution to this equation as both A_1 and H are of full rank n. Moreover $T \in L^2(C,Qd\theta)$ where Q is the spectral density of $\{x(t)\}$. This follows from $T(z) = H(z)^{-L}A_1(z)$, because $A_1 \in L^2(C,Qd\theta)$ and any left inverse of H(z) is analytic on the unit circle, in force of assumption (3.6). With this choice we get

$$S_{12}(z) = H(z)\hat{Q}(z)(T(z)^{-1})^{*}A_2(z)^{*},$$

where $\hat{Q}$ is the spectrum of $\{\hat{x}(t)\}$. From (3.5) it follows then

$$A_2(z)T(z)^{-1} = G(z)\hat{Q}(z)^{-1} .$$

Similarly, by choosing $\bar{x}(t) = \bar{T}(z)x(t)$ with $G(z)\bar{T}(z) = A_2(z)$, we obtain (3.8). In particular, for $x(t) = \hat{x}(t)$ we find $\bar{T} = \hat{Q}^{-1}$.

□

By choosing the generators as stated in Lemma 3.1, we get a unique irreducible F.A. model representative of each minimal splitting subspace X. These models, for the two different choices (3.7) and (3.8), can be written as

$$\begin{aligned} y_1(t) &= H(z)x(t) + w_1(t) , \\ y_2(t) &= G(z)Q(z)^{-1}x(t) + w_2(t) , \end{aligned} \qquad (3.10)$$

and, respectively, as

$$\begin{aligned} y_1(t) &= H(z)\bar{Q}(z)^{-1}\bar{x}(t) + w_1(t) , \\ y_2(t) &= G(z)\bar{x}(t) + w_2(t) . \end{aligned} \qquad (3.11)$$

We shall call "first" and "second" type <u>canonical forms</u> the two representations (3.10) and (3.11). Clearly each equivalence class of irreducible F.A. representations of a given spectrum S can in turn be represented by a unique 5-tuple

$$\{H,\ G\,Q^{-1},\ Q,\ R_1,\ R_2\} ,$$

or by

$$\{H\,\bar{Q}^{-1},\ G,\ \bar{Q},\ R_1,\ R_2\}\ .$$

Note that R_1 and R_2 are uniquely determined from the equalities (3.1) as functions of S_1,A_1,Q and S_2,A_2,Q. We conclude that all irreducible F.A. representations of the spectrum S, written in the first canonical form, are parametrized in a one-to-one way by the $n \times n$ nonsingular matrix function Q as

$$\{H,\ G\,Q^{-1},\ Q,\ S_1-HQH^*,\ S_2-GQ^{-1}G^*\}\ , \tag{3.12}$$

where Q is constrained to satisfy the condition that the matrix

$$\begin{bmatrix} S_1 & HG^* & HQ \\ GH^* & S_2 & G \\ QH^* & G^* & Q \end{bmatrix} \tag{3.13}$$

be a spectral density. Dually, all irreducible F.A. representations of S written in the second canonical form are parametrized in a one-to-one way by the nonsingular $n \times n$ matrix function $\bar{Q}$ as

$$\{H\bar{Q}^{-1},\ G,\ \bar{Q},\ S_1-H\bar{Q}^{-1}H^*,\ S_2-G\bar{Q}G^*\}\ , \tag{3.14}$$

where $\bar{Q}$ is constrained by the condition that the matrix

$$\begin{bmatrix} S_1 & HG^* & H \\ GH^* & S_2 & G\bar{Q} \\ H^* & \bar{Q}G^* & \bar{Q} \end{bmatrix} \tag{3.15}$$

be a spectral density function.

At this point we are ready to describe the solution set of our stochastic realization problem P.1. We introduce the $n \times n$ Hermitian matrices

$$\bar{Q}_1 := H^*S_1^{-1}H,\quad Q_2 = G^*S_2^{-1}G \tag{3.16}$$

and set

$$Q_1 := \bar{Q}_1^{-1}, \qquad \bar{Q}_2 := Q_2^{-1} . \tag{3.17}$$

Note that both $\bar{Q}_1$ and Q_2 are strictly positive definite rational spectral density matrices in force of condition (3.6) and our standing assumptions on S. We define also the $n \times n$ Hermitian matrices

$$\Delta := Q_1 - Q_2 , \quad \bar{\Delta} := \bar{Q}_2 - \bar{Q}_1 . \tag{3.18}$$

THEOREM 3.1

All irreducible F.A. representations of the spectrum S written in the first canonical form (3.12) are parametrized by the solutions Q of the matrix inequality

$$Q-Q_2-(Q-Q_2)\Delta^{-1}(Q-Q_2)^* \geq 0 . \tag{3.19}$$

Dually, all irreducible F.A. representations of S written in the second canonical form (3.14) are parametrized by the solutions $\bar{Q}$ of the inequality

$$\bar{Q}-\bar{Q}_1-(\bar{Q}-\bar{Q}_1)\bar{\Delta}^{-1}(\bar{Q}-\bar{Q}_1)^* \geq 0. \tag{3.20}$$

An $n \times n$ matrix function Q solves (3.19) if and only if $\bar{Q} = Q^{-1}$ solves (3.20). All solutions Q ($\bar{Q}$) of (3.19) (resp. (3.20)) are Hermitian bounded and strictly positive definite, in fact they satisfy

$$Q_1 \geq Q \geq Q_2 , \quad \bar{Q}_2 \geq \bar{Q} \geq \bar{Q}_1 , \tag{3.21}$$

where Q_1 and Q_2 ($\bar{Q}_1$ and $\bar{Q}_2$) are the spectral densities defined by (3.16) and (3.17).

Proof:

What needs to be shown is that an $n \times n$ matrix function Q makes (3.13) a spectral density matrix if and only if it satisfies the quadratic inequality (3.19). Assume there is a Q making (3.13) into a spectral density matrix. Then, by a standard block diagonalization procedure, the positive definiteness of (3.13) is seen to be equivalent to

$$\begin{aligned} &S_2 \geq 0 , \\ &\tilde{S}_1 := S_1 - S_{12} S_2^{-1} S_{21} \geq 0 , \\ &Q - G^* S_2^{-1} G - (Q - G^* S_2^{-1} G) H^* \tilde{S}_1^{-1} H (Q - G^* S_2^{-1} G) \geq 0 . \end{aligned} \tag{3.22}$$

The first two inequalities are trivially satisfied. In fact, by our Basic Assumption on S, S_2 and $\tilde{S}_1$ are strictly positive definite on the whole of C. By simple matrix manipulations it can be checked that

$$H^* \tilde{S}_1^{-1} H = H^* (S_1 - H Q_2 H^*)^{-1} H = (Q_1 - Q_2)^{-1} \tag{3.23}$$

and therefore, recalling our notations (3.18), we see that Q has to satisfy the inequality (3.19).

Note that Q makes the matrix (3.13) positive semidefinite if and only if $\bar{Q} = Q^{-1}$ makes (3.15) positive semidefinite. This in turn happens if and only if $\bar{Q}$ satisfies the dual inequality (3.20) as can be seen by exactly the same argument used before. Thus Q satisfies (3.19) if and only if Q^{-1} satisfies (3.20). Observe now that the matrix Δ^{-1}, given by the expression (3.23), is strictly positive definite Hermitian on the unit circle and therefore any solution Q to (3.19) makes $Q-Q_2$ positive semidefinite Hermitian. Hence Q is Hermitian and $Q \geq Q_2$. Similarly any solution $\bar{Q}$ of (3.20) satisfies $\bar{Q} \geq \bar{Q}_1$. Then, writing $\bar{Q}$ as Q^{-1}, we obtain the first inequality in (3.21). So, any solution to (3.19) has a lower (Q_2) and upper bound ($\bar{Q}_1^{-1}$), Q_2 being strictly positive definite and $\bar{Q}_1^{-1}$ being trivially bounded on C. It follows that any solution to (3.19) is a spectral density matrix. The matrix (3.13) constructed from such a solution is also Hermitian positive semidefinite and has bounded entries on the unit circle. Therefore it is a spectral density matrix.

□

The solution set of the inequalities (3.19), (3.20) can be described quite explicitly.

THEOREM 3.2

<u>An</u> $n \times n$ <u>matrix valued function</u> Q <u>on the unit circle solves the inequality</u> (3.19) <u>if and only if it is Hermitian and</u> $Q_1 \geq Q \geq Q_2$. <u>Dually, an</u> $n \times n$ <u>matrix</u> $\bar{Q}$ <u>solves</u> (3.20) <u>if and only if it is Hermitian and satisfies</u> $\bar{Q}_2 \geq \bar{Q} \geq \bar{Q}_1$.

<u>Proof</u>:

The "only if "part is already contained in the statement of Theorem 3.1. We only need to prove the "if" part. Assume first that $Q_1 > Q > Q_2$ (with strict inequalities) holds. Then (Q_1-Q) and $(Q-Q_2)$ are both Hermitian strictly positive definite and therefore $(Q-Q_2)^{-1} + (Q_1-Q)^{-1}$ is strictly positive definite. By a well-known formula for the inverse of a sum of matrices we see that this positivity condition is equivalent to

$$Q-Q_2-(Q-Q_2)\Delta^{-1}(Q-Q_2)^* > 0 . \tag{3.24}$$

Now, every Q satisfying $Q_1 \geq Q \geq Q_2$ can be approximated in $L^{\infty}_{nxn}(C)$ by a sequence of matrices Q_k for which the <u>strict</u> inequalities hold. Take for instance

$$Q_k = \frac{k-1}{k} Q + \frac{1}{2k} (Q_1+Q_2) ,$$

for which apparently $Q_k - Q_2 > 0$ and $Q_1 - Q_k > 0$. Hence Q_k satisfies the strict inequality (3.24). But the left hand side of (3.24) is a positive definite matrix which is a continuous function of Q_k and, as $k \to \infty$, it can at most become positive semidefinite.

□

REMARK

As a corollary of Theorems 3.1, 3.2 we obtain that the inequality $Q_1 \geq Q \geq Q_2$ is equivalent to

$$S_1 \geq HQH^*, \qquad S_2 \geq GQ^{-1}G^*, \qquad Q > 0 ,$$

which form in turn an equivalent set of conditions to the positivity of the matrix (3.13). This fact in particular guarantees that if Q satisfies (3.21) (or equivalently (3.19)), then the noise spectra R_1 and R_2 will be (Hermitian and) positive semidefinite. Note that the maximal solution Q_1 is in this sense just the matrix which corresponds to the largest approximant of rank n of S_1 in the ordering of Hermitian positive semidefinite matrices. □

Theorem 3.1 provides a recipe for computing all irreducible F.A. representations describing a given spectral density matrix S in a fixed coordinate system. We can now easily see that there are many of such representations (a fact that we have not bothered to show till now). For example, as the two "extreme" spectra Q_1 and Q_2 defined in (3.16) and (3.17) both satisfy the inequality (3.19) (with equality sign), we see that there are a "maximal" and "minimal" irreducible F.A. representations (in the first canonical form) which correspond respectively to the maximal (Q_1)

and minimal (Q_2) solutions to the inequality (3.19).

Solutions like Q_1, Q_2 above for which (3.19) is satisfied with equality sign have a special meaning. They correspond to joint spectra (3.13) of minimum possible rank, m, as can be seen from the block diagonalization (3.22). Since the rank of the joint spectrum of $\{z_1(t)\}$, $\{z_2(t)\}$ and $\{x(t)\}$ is equal to the multiplicity of the doubly invariant subspace $H(x,z_1,z_2)$ spanned by these processes, the multiplicity m of $H(x,z_1,z_2)$ is equal to the multiplicity of the subspace $H(z_1,z_2)$. This can only happen if $H(x,z_1,z_2) = H(z_1,z_2)$, or, that is the same, if $x(t) \in H(z_1,z_2)$ for all $t \in \mathbf{Z}$. We see that all models which correspond to solutions Q of (3.19) with equality sign are characterized by the fact that the factor process $\{x(t)\}$ is a function of $\{z_1(t)\}$, $\{z_2(t)\}$. This observation is the key to the following result.

PROPOSITION 3.1

The solutions Q to the quadratic matrix equation

$$Q-Q_2-(Q-Q_2)\Delta^{-1}(Q-Q_2)^* = 0 \tag{3.25}$$

parametrize in a one-to-one way the (strong) irreducible y-measurable representations of the processes $\{y_1(t)\}$, $\{y_2(t)\}$ of the form (3.10).

Dually, all solutions $\bar{Q}$ to the quadratic equation

$$\bar{Q} - \bar{Q}_1 - (\bar{Q}-\bar{Q}_1)\bar{\Delta}^{-1}(\bar{Q}-\bar{Q}_1)^* = 0 \tag{3.26}$$

parametrize in a one-to-one way the (strong) irreducible F.A. representations of $\{y_1(t)\}$, $\{y_2(t)\}$ of the form (3.11) for which $X \subset H(y)$.

Proof:

Consider a F.A. representation of the type (3.10). If the factor space X is contained in H(y), then $H(x,y_1,y_2) = H(y_1,y_2) = H(y)$

and hence the joint spectrum of $\{y_1(t)\}$, $\{y_2(t)\}$, $\{x(t)\}$ has rank m. This implies that the spectrum Q of $\{x(t)\}$ satisfies (3.19) with equality sign. Vice versa, assume Q is a solution of (3.25). Then, as discussed previously, the factor process of the F.A. model of type (3.3) attached to the weak realization $\{H, GQ^{-1}, Q, S_1-HQH^*, S_2-GQ^{-1}G^*\}$ of the spectrum S, has the property that x(t) belongs to $H(z_1,z_2)$ for all t. It can therefore be written as $x(t) = P_1(z)z_1(t) + P_2(z)z_2(t)$, where $P_i(z)$, $i = 1,2$, are $n \times m_i$ transfer matrices. Define an n-dimensional process $\{\tilde{x}(t)\}$ by setting

$$\tilde{x}(t) = P_1(z)y_1(t) + P_2(z)y_2(t). \tag{3.27}$$

Then $\{\tilde{x}(t)\}$, $\{y_1(t)\}$, $\{y_2(t)\}$ have exactly the same joint second order statistics (i.e. the same spectrum) as $\{x(t)\}$, $\{z_1(t)\}$, $\{z_2(t)\}$. Since conditional orthogonality depends on joint second order moments only, it then follows that $\tilde{X} := \overline{\text{span}}\,\{\tilde{x}(t);\ t \in \mathbb{Z}\}$ is splitting for $H(y_1)$, $H(y_2)$ exactly as the factor space X was splitting for $H(z_1)$, $H(z_2)$. Hence $\{\tilde{x}(t)\}$ is the factor process of a <u>strong</u> F.A. representation of the type (3.10). By construction $\{\tilde{x}(t)\}$ has spectral density matrix equal to Q and $\tilde{X} \subset H(y)$.

□

Let us define the stationary n-dimensional processes

$$\begin{aligned} x_1(t) &= Q_1(z)\,\bar{x}_1(t), \qquad \bar{x}_1(t) = H^*(z)S_1(z)^{-1}y_1(t)\,, \\ x_2(t) &= G^*(z)S_2(z)^{-1}y_2(t)\,, \end{aligned} \tag{3.28}$$

where Q_1 is defined by (3.16),(3.17). Observe that the spectra of $\{x_1(t)\}$ and $\{x_2(t)\}$ are precisely the extremal solutions Q_1,Q_2 of the quadratic inequality (3.19). It is immediate to check that $\{x_1(t)\}$ and $\{x_2(t)\}$ are minimal generators for the subspaces

$$X_1 := \bar{E}^{H(y_1)} H(y_2), \qquad X_2 := \bar{E}^{H(y_2)} H(y_1).$$

(In fact, for example X_2 is generated by $\hat{y}_1(t) = S_{12}(z) S_2^{-1}(z) y_2(t) = H(z) x_2(t)$). Moreover both X_1 and X_2 are minimal splitting subspaces (compare e.g. Lindquist, Picci and Ruckebusch, 1979)

$$X_1 \subset H(y_1) \;, \quad X_2 \subset H(y_2) \;,$$

therefore they specify two equivalence classes of strong irreducible F.A. representations of $\{y_1(t)\}$, $\{y_2(t)\}$. The particular generators $\{x_1(t)\}$ and $\{x_2(t)\}$ defined in (3.28) correspond to choosing these representations in the first canonical form, namely

$$\begin{aligned} y_1(t) &= H(z) x_1(t) + w_{1,1}(t) \;, \\ y_2(t) &= G(z) Q_1(z)^{-1} x_1(t) + w_{1,2}(t) \;, \end{aligned} \tag{3.29}$$

and

$$\begin{aligned} y_1(t) &= H(z) x_2(t) + w_{2,1}(t) \;, \\ y_2(t) &= G(z) Q_2(z)^{-1} x_2(t) + w_{2,2}(t) \;. \end{aligned} \tag{3.30}$$

Observe that in the representation (3.29) the second equation is just the decomposition of $y_2(t)$ as the sum of the (noncausal) estimate $\hat{y}_2(t) = S_{21}(z) S_1(z)^{-1} y_1(t)$ and of the corresponding estimation error. The first equation is more interesting. It can be rewritten in the form

$$y_1(t) = \Pi_H(z) y_1(t) + (I - \Pi_H(z)) y_1(t) \;, \tag{3.31}$$

where Π_H is the projection valued matrix function

$$\Pi_H(z) = H(z) (H^*(z) S_1(z)^{-1} H(z))^{-1} H^*(z) S_1(z)^{-1} \tag{3.32}$$

mapping onto the column space of H. Note that Π_H is S_1-orthogonal, i.e. $\Pi_H S_1 (I-\Pi_H)^* = 0$ a.e. on the unit circle. Thus x_1 formally looks like the classical least squares estimate of x in the linear model $y_1 = Hx + w_1$. An analogous interpretation holds for the second equation in (3.30).

The next theorem describes quite explicitly the family of all (strong) y-measurable irreducible F.A. representations of $\{y_1(t)\}$, $\{y_2(t)\}$.

THEOREM 3.3

The factor process of any irreducible y-measurable F.A. representation (in the first canonical form) is a combination of $\{x_1(t)\}$ and $\{x_2(t)\}$ of the form

$$x(t) = \Pi(z)x_1(t) + (I-\Pi(z))x_2(t) , \tag{3.33}$$

where

$$\Pi: = (Q-Q_2)\Delta^{-1} \tag{3.34}$$

is a Δ-orthogonal projection valued matrix function on the unit circle.

Proof:

The proof relies on the easily checked fact that $E[x_1(t)| |H(x_2)] = x_2(t)$. Then $\tilde{x}(t) := x_1(t) - x_2(t)$ form a process which is orthogonal to $\{x_2(t)\}$ and so $H(x_1,x_2) = H(\tilde{x}) \oplus H(x_2)$, where the direct sum is orthogonal. Now, any minimal splitting subspace $X \subset H(y)$ is actually contained in $H(x_1,x_2) \subset H(y)$ (Lindquist and Picci, 1985), so that the corresponding factor process $\{x(t)\}$ can be expressed as

$$x(t) = S_{x,\tilde{x}}(z)\Delta(z)^{-1}\tilde{x}(t) + S_{x,x_2}(z)Q_2(z)^{-1}x_2(t) , \tag{3.35}$$

where the cross spectra are easily computed from

$$S_{x,x_1} = S_{x,y_1} S_1^{-1} H Q_1 = Q \quad ,$$

$$S_{x,x_2} = S_{x,y_2} S_2^{-1} G \quad = Q_2 \quad .$$

Equation (3.35) is exactly the same as (3.33). In order to check that Π is a projection, notice that right multiplication of (3.25) by Δ^{-1} gives

$$(Q-Q_2)\Delta^{-1} = (Q-Q_2)\Delta^{-1}(Q-Q_2)\Delta^{-1} \quad ,$$

which shows that $\Pi = \Pi^2$; moreover (3.25) can be rewritten to look exactly like $\Pi\Delta(I-\Pi)^* = 0$. Thus Π is a Δ-orthogonal projection.

□

If we couple formula (3.33) with the explicit expressions (3.28) given for $x_1(t)$ and $x_2(t)$, we obtain a linear transformation acting on the "data" $\{y_1(t)\},\{y_2(t)\}$ that we want to represent. This is precisely the rule telling us how the factor process of each y-measurable representation is manifactured. Note that (3.33) is still parametrized by Q. To complete the picture we need now to describe the solution set of the quadratic equation (3.25).

PROPOSITION 3.2

Let V *be a square spectral factor of the spectral density matrix* $\Delta = Q_1 - Q_2$. *Then all solutions* $Q \neq Q_2$ *to* (3.25) *are given by*

$$Q = Q_2 + V\,\Gamma\Gamma^* V^* \quad , \tag{3.36}$$

where Γ *is any* $n \times k$ $(k \leq n)$ *isometric matrix function on the unit circle, i.e. such that*

$$\Gamma^* \Gamma = I_k \, , \tag{3.37}$$

I_k being the $k \times k$ identity matrix.

Proof:

Write $Q-Q_2$, assumed to be of rank $k \leq n$, in factorized form as

$$Q - Q_2 = U \, U^* \, ,$$

with U a full rank spectral factor of dimension $n \times k$. Since U has a left inverse, (3.25) can be reduced to

$$I_k = U^* (V \, V^*)^{-1} U \, ,$$

from which we see that $\Gamma := V^{-1} U$ satisfies (3.37).

□

Note that there is just one Q such that $\mathrm{rank}(Q-Q_2) = n$. In this case Γ is square and (3.37) is equivalent to $\Gamma\Gamma^* = I$; hence we obtain the "maximal" solution $Q = Q_1$. At the other extreme, the "minimal" solution $Q = Q_2$ is formally obtained by setting $\Gamma = 0$ in formula (3.36). Observe that by choosing Γ varying over the set of rational isometric matrices we have a parametrization of all rational solutions of equation (3.25). In other words, recalling that H and G were chosen rational, we have a parametrization of all rational irreducible y-measurable F.A. representations of the processes $\{y_1(t)\}$ and $\{y_2(t)\}$.

4. Causality

As an application of the characterization obtained in Sect. 3 we shall discuss here the question of causality of the transfer function $W(z)$ defined at the end of Sect. 2. We shall call a F.A. model

$$\begin{aligned} y_k(t) &= \hat{y}_k(t) + w_k(t) , \\ \hat{y}_k(t) &= A_k(z)x(t) , \end{aligned} \qquad k = 1,2 \tag{4.1}$$

causal whenever we can write

$$\hat{y}_2(t) = W(z)\hat{y}_1(t) \tag{4.2}$$

for a causal $m_2 \times m_1$ transfer function matrix. The question is if there are any causal F.A. models for a given pair of processes $\{y_1(t)\},\{y_2(t)\}$ satisfying our Basic Assumption. Note that for irreducible models $W(z) = A_2(z)A_1(z)^{-L}$ and at the effect of (4.2) the choice of the left inverse is immaterial. Therefore an irreducible model will be causal if, for at least one left inverse A_1^{-L}, the transfer matrix $A_2A_1^{-L}$ is causal.

We shall need the concept of Wiener-Hopf factorization relative to the unit circle C of the rational matrix $S_{12}(z)$. As noticed in (Fuhrmann and Willems, 1979), the original arguments of (Gohberg and Krein, 1960) can be adapted to cover the nonsquare (singular) case which is of interest here. Recall that by our Basic Assumption and in force of condition (3.6), $S_{12}(z)$ has constant rank n on the unit circle.

LEMMA 4.1 (WIENER-HOPF FACTORIZATION)

The rational matrix function $S_{12}(z)$ can be factored as

$$S_{12}(z) = \hat{H}(z)D(z)\hat{G}(z)^* , \tag{4.3}$$

where $\hat{H}(z)$ and $\hat{G}(z)$ are $m_1 \times n$ and $m_2 \times n$ causal rational matrices of rank n on the unit circle with a causal left inverse and $D(z)$ is an $n \times n$ diagonal matrix of the type

$$D(z) = \mathrm{diag}\{z^{-k_1},\dots,z^{-k_\ell}, z^{k_{\ell+1}},\dots,z^{k_n}\} . \tag{4.4}$$

The integers

$$-k_1 \le \dots \le -k_\ell < 0 \le k_{\ell+1} \le \dots \le k_n \tag{4.5}$$

are uniquely determined and are called the (left) Wiener-Hopf factorization indices of S_{12}, relative to C .

□

Note that $D(z)$ can in turn be factored as

$$D(z) = D_1(z)D_2(z)^* , \tag{4.6}$$

where

$$\begin{aligned} D_1(z) &= \mathrm{diag}\ \{z^{-k_1},\dots,z^{-k_\ell}, 1,\dots,1\} , \\ D_2(z) &= \mathrm{diag}\ \{1,\dots,1, z^{-k_{\ell+1}},\dots,z^{-k_n}\} . \end{aligned} \tag{4.7}$$

The factorization (4.3) can thus be rewritten in the form

$$S_{12}(z) = [\hat{H}(z)D_1(z)]\,[\hat{G}(z)D_2(z)]^* . \tag{4.8}$$

In this section we shall identify the rational matrix functions $H(z)$ and $G(z)$ of the minimal factorization (3.5) with the two terms within square brackets in (4.8). We shall consider irreducible F.A. models written in the second canonical form

$$y_1(t) = \hat{H}(z)D_1(z)\bar{Q}(z)^{-1}\bar{x}(t) + w_1(t) ,$$
$$y_2(t) = \hat{G}(z)D_2(z)\bar{x}(t) \qquad + w_2(t) , \tag{4.9}$$

with $\bar{Q}(z)$ any Hermitian $n \times n$ matrix function satisfying

$$\bar{Q}_2 \geq \bar{Q} \geq \bar{Q}_1, \tag{4.10}$$

where

$$\bar{Q}_2 = (D_2^* \hat{G}^* S_2^{-1} \hat{G}\, D_2)^{-1} , \tag{4.11}$$

$$\bar{Q}_1 = D_1^* \hat{H}^* S_1^{-1} \hat{H}\, D_1 . \tag{4.12}$$

In this framework the transfer function matrix W relative to an arbitrary irreducible F.A. model is

$$W(z) = \hat{G}(z)D_2(z)\bar{Q}(z)D_1(z)^* \hat{H}(z)^{-L} . \tag{4.13}$$

LEMMA 4.2

<u>The transfer function</u> $W(z)$ <u>is causal (for at least one choice of the left inverse</u> $\hat{H}^{-L}$<u>) if and only if</u> $D_2(z)\bar{Q}(z)D_1(z)^*$ <u>is a causal matrix function.</u>

<u>Proof</u>:

(If). Since $\hat{H}(z)$ is minimum phase there is a causal left inverse. Thus if $D_2\bar{Q}\,D_1^*$ is causal, W is causal.

(Only if). Since

$$D_2(z)\bar{Q}(z)D_1(z)^* = \hat{G}(z)^{-L}W(z)\hat{H}(z)$$

and $\hat{G}(z)$ is minimum phase, it follows that W causal implies $D_2\bar{Q}\,D_1^*$ causal.

□

THEOREM 4.1

Under the stated assumptions a causal irreducible F.A. model of $\{y_1(t)\}$, $\{y_2(t)\}$ can only exist if the Wiener-Hopf factorization indices of $S_{12}(z)$ are all nonnegative (i.e. $D_1(z) = I_n$ in (4.7)).

Proof:

We show that if $D_1(z) \neq I_n$ or equivalently $\ell > 0$ in (4.5), then $D_2 \bar{Q} D_1^*$ cannot be causal. In fact, for $j \leq \ell$, the j-th diagonal element of $D_2 \bar{Q} D_1^*$ is $z^{k_j} \bar{q}_{jj}(z)$, where $\bar{q}_{jj}(z)$ is the j-th diagonal element of $\bar{Q}(z)$. By definition of causality we must have (compare (1.9))

$$\int_{-\pi}^{\pi} e^{i\theta(k_j-k)} \bar{q}_{jj}(e^{i\theta}) d\theta/2\pi = 0$$

for all $k > 0$. By taking complex conjugate and recalling that $\bar{q}_{jj}$ is a real function, we also obtain

$$\int_{-\pi}^{\pi} e^{i\theta(k-k_j)} \bar{q}_{jj}(e^{i\theta}) d\theta/2\pi = 0$$

for $k > 0$. Now, if these two relations hold for some $k_j > 0$, they imply that

$$\int_{-\pi}^{\pi} e^{-i\theta h} \bar{q}_{jj}(e^{i\theta}) d\theta/2\pi = 0$$

for all $h \in \mathbf{Z}$. This is equivalent to $\bar{q}_{jj}(e^{i\theta}) = 0$ a.e. on C and contradicts the (strict) positive definiteness of $\bar{Q}(e^{i\theta})$. Thus $D_2 \bar{Q} D_1^*$ cannot be causal.

□

At this point, to be able to proceed any further we have to introduce the assumption that the Wiener-Hopf indices of S_{12} are all nonnegative, i.e. that

$$D(z) = D_2^*(z) = \mathrm{diag}\{z^{k_1},\dots,z^{k_n}\} , \tag{4.14}$$

where

$$0 \le k_1 \le \dots \le k_n . \tag{4.15}$$

We next introduce the notion of $n \times n$ <u>trigonometric polynomial matrix</u> $P(z) = [p_{ij}(z)]$ with <u>indices</u> the ordered set of n natural numbers $\{k_1,\dots,k_n\}$. The i,j-th entry of $P(z)$ has the structure

$$p_{ij}(z) = \sum_{-k_j}^{k_i} p_{ijk} z^k . \tag{4.16}$$

THEOREM 4.2

<u>Assume that the Wiener-Hopf factorization indices of</u> $S_{12}(z)$ <u>are all nonnegative. Then there are causal irreducible F.A. models if and only if there are Hermitian trigonometric polynomial solutions</u> $\bar{Q}$ <u>to the inequality</u>

$$(D_2^* \hat{G}^* S_2^{-1} \hat{G} D_2)^{-1} \ge \bar{Q} \ge \hat{H}^* S_1^{-1} \hat{H} , \tag{4.17}$$

<u>with indices equal to the factorization indices</u> (4.15) <u>of</u> $S_{12}(z)$.

<u>Proof:</u>

By our assumption (4.14) and Lemma 4.2 each causal F.A. model is characterized by $D_2(z)\bar{Q}(z)$ being causal. By definition this happens if and only if

$$\int_{-\pi}^{\pi} e^{-i\theta(k_i+k)} \bar{q}_{ij}(e^{i\theta})\, d\theta/2\pi = 0$$

for $i,j = 1,\ldots,n$ and all $k > 0$. By taking complex conjugate and recalling that $\bar{Q}$ is Hermitian, i.e. $\bar{q}_{ij}(e^{i\theta})^* = \bar{q}_{ji}(e^{i\theta})$, we also get

$$\int_{-\pi}^{\pi} e^{i\theta(k_j+k)}\bar{q}_{ij}(e^{i\theta})d\theta/2\pi = 0$$

for all $k > 0$. Taken together these two relations are equivalent to

$$\int_{-\pi}^{\pi} e^{-i\theta h}\bar{q}_{ij}(e^{i\theta})d\theta/2\pi = 0$$

for all h satisfying $h < -k_j$ and $h > k_i$. This shows that $\bar{q}_{ij}(z)$ has the expression (4.16).

□

Observe that a positive definite trigonometric polynomial $\bar{Q}$ can be factored as

$$\bar{Q}(z) = \bar{N}(z)\bar{N}(z)^* , \tag{4.18}$$

where $\bar{N}(z)$ is an $n \times n$ polynomial matrix which can be taken row-proper and with row degrees exactly equal to the indices $k_1 \leq \ldots \leq k_n$ of $\bar{Q}$. Recalling the remark made after the proof of Theorem 3.2, we can recast the conditions of Theorem 4.2 in terms of the joint spectrum S in the following way.

COROLLARY 4.1

Assume the Wiener-Hopf factorization indices of $S_{12}(z)$ are all nonnegative. Then there are causal irreducible F.A. models if and only if there are $n \times n$ polynomial matrices $\bar{N}(z)$ with (ordered) row degrees equal to the indices $k_1 \leq \ldots \leq k_n$ of $S_{12}(z)$ such that

$$S_2 \geq \hat{G} D_2 \bar{N} \bar{N}^* D_2^* \hat{G}^* ,$$
$$S_1 \geq \hat{H} (\bar{N}^*)^{-1} \bar{N}^{-1} \hat{H}^* . \tag{4.19}$$

□

At the beginning of this section causality was defined with respect to a certain choice of input $(\hat{y}_1)$ and output $(\hat{y}_2)$ variables. If we choose instead $\hat{y}_2(t)$ as input and $\hat{y}_1(t)$ as output , we can of course go through a very similar analysis and obtain analogous conditions for the existence of causal irreducible F.A. models of the type

$$\begin{aligned} y_1(t) &= W(z)^{\#} \hat{y}_2(t) + w_1(t), \\ y_2(t) &= \hat{y}_2(t) \qquad\qquad + w_2(t), \end{aligned} \tag{4.20}$$

where $\hat{y}_1(t)$ is obtained as

$$\hat{y}_1(t) = W(z)^{\#} \hat{y}_2(t) \tag{4.21}$$

for a causal $m_1 \times m_2$ transfer function matrix. Relative to the Wiener-Hopf factorization (4.3), $W^{\#}$ has the expression

$$W(z)^{\#} = \hat{H}(z) D_1(z) Q(z) D_2(z)^* \hat{G}(z)^{-L} . \tag{4.22}$$

THEOREM 4.3

Causal irreducible F.A. models of the type (4.20) can only exist if the Wiener-Hopf factorization indices of $S_{12}(z)$ are all negative or zero, i.e. only if

$$D(z) = D_1(z) = \text{diag}\{z^{-k_1}, \ldots, z^{-k_n}\} . \tag{4.23}$$

In case (4.23) is satisfied, there are causal irreducible F.A. models of the type (4.20) if and only if there are Hermitian trigonometric polynomial solutions Q to the inequality

$$(D_1^* \hat{H}^* S_1^{-1} \hat{H} D_1)^{-1} \geq Q \geq \hat{G}^* S_2^{-1} \hat{G} \ , \qquad (4.24)$$

with indices equal to the opposite $\{k_1, \ldots, k_n\}$ of the factorization indices of $S_{12}(z)$. An equivalent condition is the existence of $n \times n$ polynomial matrices $N(z)$ with (ordered) row degrees $k_1, \ldots, k_n$ such that

$$\begin{aligned} S_1 &\geq \hat{H} D_1 N N^* D_1^* \hat{H}^* \ , \\ S_2 &\geq \hat{G} (N^*)^{-1} N^{-1} \hat{G}^* \ . \end{aligned} \qquad (4.25)$$

□

Let us agree to call minimum phase those F.A. models for which both (4.2) and (4.21) are causal input-output relations. Then, as a corollary of Theorems 4.1 ÷ 4.3, we get that minimum phase models exist only if the Wiener-Hopf factorization of $S_{12}(z)$ has $D(z) = I_n$ (i.e. is "canonical" in the terminology of Gohberg and Krein (1960) and Bart, Gohberg and Kaashoek (1979)). There exist minimum phase F.A. models if and only if there are constant Hermitian n x n matrices Q for which

$$\begin{aligned} S_1 &\geq \hat{H} Q \hat{H}^* \ , \\ S_2 &\geq \hat{G} Q^{-1} \hat{G}^* \end{aligned} \qquad (4.26)$$

on the unit circle. We see that the factor process of minimum phase irreducible F.A. models written in either canonical form, for which H and G are chosen equal to the Wiener-Hopf factors, must be an n-dimensional white noise process.

References

Anderson, B.D.O. (1985): Identification of Scalar Errors-In-Variables Models with Dynamics. Automatica, 21, 709-716.

Anderson, B.D.O. and M. Deistler (1984): Identifiability in Dynamic Errors-In-Variables Models. J. Time Series Analysis, 5, 1-13.

Anderson, B.D.O. and M.R. Gevers (1982): Identifiability of Linear Stochastic Systems Operating Under Linear Feedback. Automatica, 18, 195-213.

Bart, H., I. Gohberg and M.A. Kaashoek (1979): Minimal Factorization of Matrix and Operator Functions, Operator Theory: Advances and Applications, Vol. 1, Birkhäuser Verlag, Basel.

Bart, H., I. Gohberg and M.A. Kaashoek (1984): Wiener-Hopf Factorization and Realization, in Proc. Int. Symp. on Mathematical Theory of Networks and Systems, Beer Sheva, Israel, June 1983, Springer-Verlag Lect. Notes in Control and Inf. Sciences, 58, 42-62.

Caines, P.E. and C.W. Chan (1975): Feedback Between Stationary Stochastic Processes. IEEE Trans. Aut. Control, AC-20, 498-508.

Caines, P.E. and C.W. Chan (1976): Estimation, Identification and Feedback, in System Identification: Advances and Case Studies, R.K. Mehra and D.G. Lainiotis eds., Academic Press, New York.

Deistler, M. (1985): Identifiability and Causality in Linear Dynamic Errors-In-Variables Systems. Report, Inst. of Econometrics and Operations Research, University of Technology,Vienna.

Finesso, L. and G. Picci (1984): Linear Statistical Models and Stochastic Realization Theory, in Proc. VI-th Int. Conf. on Analysis and Optimization of Systems, Nice, France, June 1984, Springer-Verlag Lect. Notes in Control and Inf. Sciences, 62, 445-470.

Fuhrmann, P.A. (1981): Linear Operators and Systems in Hilbert Space, McGraw-Hill, New York.

Fuhrmann, P.A. and J.C. Willems (1979): Factorization Indices at Infinity for Rational Matrix Functions. Integral Equations and Operator Theory, 2, 287-301.

Gevers, M.R. and B.D.O. Anderson (1981): Representations of Jointly Stationary Stochastic Feedback Processes. Int. J. of Control, 33, 777-809.

Gevers, M.R. and B.D.O. Anderson (1982): On Jointly Stationary Feedback-Free Stochastic Processes. IEEE Trans. Aut. Control, AC-27, 431-436.

Gohberg, I. and M.G. Krein (1960): Systems of Integral Equations on a Half Line with Kernels Depending on the Difference of Arguments. Amer. Math. Soc. Transl.(2), 14, 217-287.

Granger, C.W.J. (1963): Economic Processes Involving Feedback. Information and Control, 6, 28-48.

Granger, C.W.J. (1969): Investigating Causal Relations by Econometric Models and Cross-Spectral Methods. Econometrica, 37.

Hoffman, K. (1962): Banach Spaces and Analytic Functions, Prentice-Hall, Englewood Cliffs.

Kalman, R.E.(1982a): System Identification from Noisy Data, in Dynamical Systems II, A.R. Bednarek and L. Cesari eds., Academic Press, New York.

Kalman, R.E. (1982b): Identification from Real Data, in Current Developments in the Interface: Economics, Econometrics, Mathematics, M. Hazewinkel and A.H.G. Rinnooy Kan eds., Reidel, Dordrecht.

Kalman, R.E. (1983): Identifiability and Modeling in Econometrics, in Developments in Statistics, Vol. 4, P.R. Krishnaiah ed., Academic Press, New York.

Lindquist, A. and G. Picci (1985): Realization Theory for Multivariate Stationary Gaussian Processes. SIAM J. Control and Optim. 23, 809-857.

Lindquist, A., G. Picci and G. Ruckebusch (1979): On Minimal Splitting Subspaces and Markovian Representations. Math. Systems Theory, 12, 271-279.

Picci, G. and S. Pinzoni (1986): Dynamic Factor Analysis Models for Stationary Processes, IMA J. Math. Control and Information, to appear.

Rozanov, Y.A. (1967): Stationary Random Processes, Holden-Day, San Francisco.

Ruckebusch, G. (1976): Représentations Markoviennes de Processus Gaussiens Stationnaires. C.R. Acad. Sc. Paris, Sér. A, 282, 649-651.

Van Schuppen, J.H. (1985): Stochastic Realization Problems Motivated by Econometric Modelling. Report OS-R8507, Centre for Mathematics and Computer Science, Amsterdam.

Willems, J.C. (1979): System Theoretic Models for the Analysis of Physical Systems. Ricerche di Automatica, 10, 71-106.

Chapter 4

Predictive and Nonpredictive Minimum Description Length Principles

Jorma Rissanen

1. Introduction

Statistical estimation or modeling is an activity aimed at infering from a set of observed data certain properties that are expected to hold in future data. This involves a fundamental dilemma in that whatever we estimate will be determined by the current data, and yet the success of our attempts will be judged by the behaviour in the future data, which evidently are not available now. The way this difficulty is dealt with in traditional statistics is to regard the current data as a sample from a larger, in effect an infinite population, represented by a "true" probability distribution with parameters, each meant to define some property of the data. These parameters then provide the targets to be estimated, which can be done by minimization of some measure of nearness, such as the squared deviations or the likelihood function, between the existing data and the fitted parametric distributions. In trivial cases the number of the "true" parameters is taken to be known, but frequently, in order to leave all the doors open, the "true" parent distribution is assumed to have infinitely many parameters, which evidently is a safe hypothesis in that it can neither be verified nor disproved.

The problem with the "true" distribution hypothesis is not so much the fact that the distribution has to be chosen subjectively (in fact, selecting a large enough class will allow a lot of leeway) as the fact that this hypothesis forces us to regard models as approximations of the assumed distribution, the goodness of which, however, must be judged in the light of the observed data. Hence, if we fit models having different numbers of parameters, then a model with more parameters is likely to provide a better fit than one with fewer parameters without any guarantee of better performance on future data. And this is true no matter how we measure the nearness. What instead is needed is the ability to compare models regardless of the number of parameters they have, which simply cannot be done by their nearness to an abstract and subjectively selected parent distribution.

In this paper we present in a tutorial fashion a rather different approach to statistical reasoning, introduced and studied in a number of papers, Rissanen (1978), (1983b), (1984a,b,c), (1985a,b). The reasoning goes as follows: The main problem in statistical modeling is regarded as one of understanding and explaining the set of observed data, which, to be sure, often look quite chaotic. Intuitively, "understanding" presumably means something related to an ability to learn and to discover various regular features that constrain the data and that imply redundancy if we were to describe the data without taking them into account. Additionally, an understanding permits a degree of prediction. A trivial example is a sequence such as 1, 4, 9, 16, 25, which, if we spot the rule, can be described very concisely as well as perfectly predicted, provided, of course, that the rule holds even in the future. A less trivial example is Newton's law of gravitation, which is a model that permitted a great improvement over the Ptolemaic models by Eudoxus and Hipparchus as well as Tycho Brahe's tables for the planetary motions both in regard of the description length and predictability. Notice, in particular, that no "true" law is needed to do prediction; for example, Hipparchus' epicycles and eccentric circles were clearly incorrect explanations of the planetary motions, but still they provided useful predictions of the lunar eclipses. Similarly, Newton's law is also incorrect, but it provides predictions with astonishing accuracy. Many people think that there is a difference of a kind between a grossly incorrect model like Eudoxus' and an accurate one like Newton's, because, indeed, Newton's model "explains" the planetary motions with help of the most elegant law of universal gravitation. But the difference is simply one of degree, and the universal "law" of gravitation is just another incorrect model, which, incidentally, involves the rather disturbing and, in fact, absurd idea that a force is being transmitted instantly. To summarize, there are no "true" laws nor systems outside the realm of mathematics, but that does not prevent us from understanding observed data.

We are interested in statistical features which, of course, somehow reflect the underlying data generating machinery. Since we usually are not allowed to open up the machinery and take a direct look, we must have some means to recognize the regular features in the observations and to measure their amount. This can be done by counting the number of binary digits with which the observed data can be written down by taking advantage of the various models, that serve as an expression of the rules. In technical terms, we say that the data is encoded for the purpose of getting a short code length; i.e., "compressed". The resulting code length, then, represents a universal and immutable criterion for model fitting, which is just about as free from subjective and whimsical choices as we can make it. There remains the subjective selection of the class of models, but that must necessarily be so; after all, how can we learn from the data unless we can formulate the properties we wish to find? Similarly, by carefully selecting the model class we can also influence the properties we wish to discover, which gives us a means of learning. It is important to see that the code must include the description of the model itself, for otherwise the imagined decoder could not recover the data. We call the process of minimizing such a criterion the *Minimum Description Length* (*MDL*) principle.

It is clear that the length of coding the data cannot be reduced below a certain level, which is entirely determined by the data and the class of the selected models, regardless of whether the models have the same number of parameters or not. We call this critical level the *stochastic complexity* of the data, relative to the considered class of models. Different model classes can be judged by their stochastic complexity, and perhaps to some consternation a subclass of another class may produce a strictly smaller stochastic complexity than the larger class. Hence, the way to good models is not just to make the model classes larger and larger; that is to say, to make the models increasingly complex. In a stark contrast with the traditional statistics, the optimal model, determined by the stochastic complexity, or with which it is reached, is not an approximation of anything at all. Rather, it has an independent meaning in incorporating all the statistical information in the data that can be extracted with the considered class of models. In particular, it has an optimal number of parameters, which are calculated by an estimator which is either efficient or it approaches an efficient estimator in the traditional sense. In addition, stochastic complexity also sets the greatest lower bound with which the data can be predicted with the considered class of models, and we may say that its calculation and the search for a class of models giving a small stochastic complexity are the two fundamental problems in statistics.

It may be worthwhile to elaborate our view on modeling a bit further. Quite often the observed sequence is regarded as a random sample, and it is thought to consist of the important information bearing part and of the random noise that is clearly a nuisance to be gotten rid of. Hence, one may think that it is necessary to extract somehow the "useful" signal so that we then can fit our models to it. Such a prefiltering, however, hides a dangerous prejudice, for strictly speaking the observations never include any noise. They are just numbers, and the only way to separate a portion off them and to call it noise is to use models. Hence, noise is something we define it to be, namely, the difference between a modeled signal and the observed numbers, rather than something imposed by nature. The fact is that nature produces the observations, and the rest is man made - to paraphrase the famous saying of Kronecker. To take a simple example, we may model the observed input u and the observed output y as being related as follows,

$$x_t = f(u_t)$$
$$y_t = x_t + e_t ,$$

where x_t is considered the "useful" signal and e_t represents the "noise". Evidently, for a given pair of the observations this decomposition depends completely on the modeled function f. We may, of course, impose a condition on the non-observed e_t, such as that its variance is a prescribed number. The effect is a restriction on the functions f that satisfy the extra requirement, which is perfectly in order. A superficial thinking might lead one to the idea that the *MDL* principle, which has no prefilters for noise, forces us to fit models to noise. Such thinking is evidently contradictory, because as we just saw, noise itself is a result of the modeling process. The *MDL* principle fits models to data, and we can actually see directly how it automatically avoids inserting parameters to capture "noise": If, indeed, a certain portion of the observations consists of random fluctuations, such as e_t in the previous example, then no modeling can shorten their description; i.e., to "compress" them. Suppose that, say, two parameters in the modeled function f are sufficient to compress the "useful" part in the observations and that we try to add parameters to compress the fluctuations. Since these cannot be compressed by any means whatsoever, the extra parameters do not "buy" any compression while their own description costs bits. Hence, the *MDL* principle will remove such parameters, and what remains are only the effective ones, which is just what is needed. The random fluctuations just "pass through" the model unchanged, and they do no harm. As a matter of fact, to push this point to its extreme, you can add pure random noise to the observations without much effect to the optimal *MDL* model!

A deeper issue involves the question of how to measure the amount of "information", that we call complexity, in the data. Fisher's famous idea was to measure this information content by the determinant of his information matrix, which by Cramer - Rao inequality represents the smallest variance of the parameter estimates. Hence, intuitively, if an estimator does achieve this lower bound, then it must be the case that the process has extracted all the useful information in the data. This is a curious, roundabout procedure, and it works just because the considered parametric likelihood function is restricted to have a fixed number of parameters. When we consider the larger classes of models where the number of parameters is not fixed, as we must in order to get better models, then the variance of the parameter estimates ceases to be a meaningful measure of the information content in the data, which is why we define the stochastic complexity directly in terms of the data, "noise" and all. In a sense, then, Fisher's idea is brilliant but the concept is far too restricted to do what was intended. The information matrix still, of course, is an important quantity, but not as a measure of the useful information in the data.

As a further point, in our view in the absence of prior knowledge we must take every observed sequence to be "typical" in that it is representative of the underlying mechanism. After all, it is all we have, and we have no right to claim otherwise. Hence, we should fit models to these data and not to what we might imagine the data should be. Only if we have a rather firm idea of the probabilistic model of a source, such as the gambling machines, obtained on some prior grounds can we claim that a certain odd observation sequence might not be typical. As a final point, it is often felt that a model is good if its parameters are such that repeated estimates are close to each other; in other words, their estimated variance is small. This is the thinking of confidence intervals and such. Well, let me define a one- parameter model, where the parameter has the value 1.2 no matter what the data are. Clearly, you cannot have a smaller variance, but such a model is probably worthless. In fact, it may be a very dangerous prejudice to isolate a quantity and call it a parameter. An example is the blood pressure, which has been regarded as an important "parameter", carrying information about the health of a human body. It has been found relatively recently that it should be regarded as a variable, because it fluctuates considerably in perfectly healthy people, and to diagnose illness, because a measurement happens to deviate from what has been thought to be its normal value, has led to needless and even dangerous medications. In conclusion, the moral of all this discussion is that it is important to understand the nature of statistical reasoning and modeling in order to be able to avoid the many pitfalls that lurk along the way, the most important of which are the arbitrary and unjustified choices that have a tendency of creeping in even if we are on our guard. The only reality consists of the data; the rest are models and other theories, which well may be gray - as Goethe claimed - but which we ought to select as intelligently as we can.

The stochastic complexity clearly has its roots in the algorithmic notion of information, Solomonoff (1964), Kolmogorov (1965), and Chaitin (1975), which defines the complexity of a binary string to be the length of the shortest program needed to generate it in a universal computer. However, in order to make the principle practicable we must not select the class of models too rich - certainly not to include all computable functions - because then the complexity can neither be computed nor estimated by any algorithm.

2. Coding and Prediction

Our modeling principle is founded on the issues of how to describe or *encode* data efficiently; that is, with short code length. Although we do not really need any details of such codes, it nevertheless is useful to have an idea of the relevant issues in coding, above all, the code length. This may also add perspective to those interested in prediction, for the reason that it turns out to be a special case of coding. We begin with the traditional coding problem involving one probability distribution, and then we discuss the newer more general situation involving a family of them. We consider the observed data to be a string of symbols $y = y_1, \ldots, y_n$, each symbol, for instance, being a binary number written with some finite precision. We do not need to specify this precision, and, in fact, the reader may think of these observations to be just numbers as usual. More generally, the data string may consist of pairs (u_i, y_i), where the first component is an observed input and the second the observed output response. Nothing substantially different arises from this generalization; instead of a distribution for the outputs we simply consider a conditional distribution for the outputs given the inputs. A code C, then, is a one-to-one function taking each string y of every length n to a *binary* string $C(y)$. Moreover, the code length $L(y)$, defined to be the number of binary digits in $C(y)$, is required to satisfy the so-called Kraft inequality, Abramson (1968),

$$\sum_{y \in Y^n} 2^{-L(y)} \leq 1, \tag{2.1}$$

for all n, where Y^n denotes the set of all strings of length n. This inequality is intimately connected with a desirable property of the code, known as the prefix property, which means that no code string $C(y)$ is a prefix of another $C(y')$, where y and y' are two distinct strings of the same length. If we place the code strings $C(y)$, y running through the set of all strings, in a binary tree (in an obvious fashion), then each code string appears as a leaf having no successor nodes. But then, given such a tree, we can tell which initial portion in any string of binary symbols defines a valid code string. In other words, we can tell without a comma when we have reached the end of a code string. It is not an accident that with such a self containing description of data the code length defines a distribution $Q(y) = 2^{-L(y)}$ in Y^n if we just add the requirement that the code be efficient in the sense that the code strings have no superfluous digits, which turns (2.1) into an equality.

Suppose the strings in Y^n have a probability distribution $P(y)$ assigned to them. Then, for any code with the length satisfying (2.1), we get by Jensen's inequality, stating that for a convex function $f(x)$, $Ef(x) \leq f(Ex)$,

$$E \log \frac{2^{-L(y)}}{P(y)} = - E\, L(y) - E \log P(y) \leq \log \sum_{y \in Y^n} P(y) 2^{-L(y)} \leq 0,$$

where the equality holds if and only if $2^{-L(y)} = P(y)$ for all y. In other words, the mean code length satisfies the inequality due to Shannon: $EL(y) \geq H(n)$, where $H(n) = - \sum_{y \in Y^n} P(y) \log P(y)$ denotes the entropy of the strings of length n. This means that the ideal way to encode the strings relative to the given distribution is to assign to string y a code string with length $-\log P(y)$. This, of course, cannot always be done exactly because a code string must have an integer length, but at least we know what we should be striving for, and we call it the *ideal* code length. Another good name would be Shannon *complexity* of y relative to the given distribution.

A direct application of these ideas to compressing strings confronts us with the same problem as met in traditional statistics: The distribution $P(y)$ is not known to us, and it either has to be imagined or, better, estimated. For this reason we consider a parametric family $\{ P_\theta(y) \}$ of such distributions or *models*, where $\theta = (\theta_1, \ldots, \theta_k)$, and k ranges over the set of all natural numbers. How now to calculate the ideal code length is the central problem in the *MDL* principle to be discussed next.

There are two basic ways to go about encoding a string of data. In the first way we read the entire string and we somehow form the best estimate $\hat{\theta}(y)$ of the parameter vector θ. Then we design a code C such that the length of the code string $C(y)$ is close to the ideal $-\log P_{\hat{\theta}(y)}(y)$. We need not concern ourselves with the details of how such a code can be designed, which is just a routine matter. The important thing to realize is that the data y can be decoded from the code string $C(y)$ only if the decoder also knows the estimated parameter vector $\hat{\theta}(y)$. This has to be given in an explicitly coded form, because the decoder at the time it is needed does not yet know y and, hence, cannot calculate the estimate by any conceivable algorithm. The binary code string for the parameter vector, which may be placed as a preamble in front of $C(y)$, must clearly be a prefix code, for otherwise the decoder would not be able to separate it from the subsequent binary code of the data. Hence, its length $L(\theta)$ must satisfy the Kraft-inequality, $\Sigma 2^{-L(\theta)} \leq 1$, where θ runs through all its possible values. These values are clearly truncations (think of computing the maximum likelihood estimates, which surely result in truncated numbers). If we carry too many fractional digits, the required code will have to be long, while if we truncate too heavily, the results will deviate too much from the optimum, and we end up coding the string with non-optimal parameters. It turns out that when each component is truncated to its optimal precision, reflecting its importance to the entire code length, the code length for the k-component parameter vector and the loss due to truncation is $\frac{k}{2} \log n$ bits, Rissanen (1978). In addition, the decoder will have to be given the number of the components k in the estimated parameter vector as another prefix coded preamble, which takes a little more than $\log k$ bits. This number, of course, is almost always quite negligible in comparison with the other length, and we drop it. All told, the best ideal code length with this type of "nonpredictive" coding is to within terms of order $\log n$ given by

$$I_{NP}(y) = \min_{k,\theta} \{ -\log P_\theta(y) + \frac{k}{2} \log n \}. \tag{2.2}$$

The same expression but with different content and scope was also derived by restrictive Bayesian assumptions in Schwarz (1978). We also refer to the pioneering work of Akaike (1974) for another criterion, where the weaker model complexity penalizing term k gets added to the first, the negative logarithm of the likelihood term. In contrast with (2.2) such a term is too weak to produce consistent estimates of the number of parameters in all the analyzed cases, Hannan (1980) and Shibata (1976). Finally, we add that when the parameter coding job is done more carefully, Rissanen (1983a), a third term is required, namely, $k \log \|\theta\|_{M(\theta)}$, where $M(\theta)$ denotes the Hessian matrix of $-\log P_\theta(y)$. This term turns out to be sensitive to the structure in which the parameters of a multivariable dynamic system are represented, Rissanen (1983b); see also Section 4.

The other way of coding data strings requires no explicit code for the parameters, because the coding will be done in a "predictive" way. What this means is that from each portion $y^t = y_1, \ldots, y_t$ of the data string we form an estimate of the distribution $P_\theta(y_{t+1} \mid y^t)$ for the possible values of the next symbol, where θ is to be replaced by an estimated value $\hat{\theta}(t) = \hat{\theta}(y^t)$, calculated by an algorithm from the so-far processed string. The decoder knows this algorithm, and he can also calculate the same estimate provided that it indeed does not depend on the future and not yet decoded data points. We know from Shannon's result, derived above, that the best way to do the coding is to assign to the next symbol the code length $-\log P_{\hat{\theta}(t)}(y_{t+1} \mid y^t)$, and hence the best total ideal code length with this type of predictive coding is

$$I_P(y) = \min_k \{ -\sum_{t=0}^{n-1} \log P_{\hat{\theta}(t)}(y_{t+1} \mid y^t)\}. \tag{2.3}$$

We should also have included the code length, $\log k$, required to describe the number of the components in the estimated parameter vectors, but as above this term is negligible. How should we pick the estimates $\hat{\theta}(t)$? It seems to make eminent sense to pick them in such a way that the accumulated past code lengths

$$-\log P_\theta(y^t) = -\sum_{t=0}^{t-1} \log P_\theta(y_{t+1} \mid y^t), \tag{2.4}$$

are minimized, which is seen to be done by the maximum likelihood estimates of the parameters for each value of k. This represents a most attractive principle of inductive inference: Make that choice that has worked best in the past. And who can argue against that, provided that we have no other "prior" knowledge about the behavior of the data! This philosophy in his "prequential" approach to estimation was also discovered independently in Dawid (1984). A somewhat similar and yet crucially different "cross- validation" principle has been studied in Stone (1977). Because no "honesty" of the predictions is required, the associated criterion is asymptotically equivalent with Akaike's AIC, and hence the resulting estimates of the number of parameters are not consistent.

In order to avoid ill-conditioned optimization problems, we in (2.4) never estimate more parameters than data points; that is, we begin with $k = 0$ and increase k gradually to each final value with which the criterion in (2.3) is evaluated. The case with no free parameters means that we need an initial distribution $P(y_1)$ to predict or encode the very first observation. This could be done by having a fixed parameter value $\theta(0)$, obtained somehow on prior grounds, which singles out a distribution from the family. We discuss later how such a prior knowledge can be taken advantage of in modeling and prediction.

The predictive coding process is seen to be very similar to prediction: In both cases we try to unravel the uncertainty about the "next" observation y_{t+1} by acting on the past data only. In fact, the two processes are equivalent. To see this, let $\delta(y_{t+1} - \hat{y}(t+1 \mid t))$ be any reasonable prediction error measure, where $\hat{y}(t+1 \mid t)$ is some prediction of the next observation, involving parameters to be estimated from the past data. Define a conditional density $f_\theta(y_{t+1} \mid y^t)$ proportional to $e^{-\delta(y_{t+1} - \hat{y}(t+1 \mid t))}$, and we get a family of parametric probabilistic models, where the code length, apart from an irrelevant term due to truncation and proportional to n, is the sum of the prediction errors. A particularly important special case results from the quadratic prediction error measure, because then the predictive *MDL* principle reduces to a predictive least squares (*LS*) principle. We discuss its application to ARMA estimation in the next section. Because the non-predictive coding process cannot be interpreted as prediction, we conclude that coding is a strictly more general process than prediction.

We conclude this section by stating that the two described coding lengths are asymptotically optimal in the sense that their mean, relative to any process in the considered class of "smooth" models, is shortest among all codes satisfying (2.1). Because the variance of these lengths, computed per observation, behaves like $1/n$, we may take these lengths themselves to represent well the shortest possible per symbol code lengths (prediction errors), and we call $I_{NP}(y)$ and $I_P(y)$ the non-predictive and predictive *stochastic complexities*, respectively, of the string y, relative to the considered class of models. This result not only generalizes the above mentioned Shannon theorem, giving a tight lower bound for the code length and the prediction errors, but it also serves a similar role as Cramer-Rao inequality for estimators, except that we may assess the goodness of any estimators, including the number of parameters. The name "complexity" seems apt in view of the fact that it represents the ultimate limit to which the three fundamental tasks, prediction, estimation, and coding, can be performed.

3. ARMA Estimation and Prediction

As we outlined in the preceding section, estimation and prediction are intertwined: you cannot predict optimally without performing estimation optimally. Here we mean the real prediction problem where we are given an observed sequence of numbers, $y(1), \dots, y(n)$, one by one, and we are asked to predict for each n the next value. This is to be done without knowing the probabilistic source of the numbers as usually done in prediction theory. Our approach is to select a class of models, or perhaps several classes, and fit a model in each class with the predictive *LS* principle. The prediction will be done with the best model at each instant of time, and if the past is any guidance to the future this strategy will provide the best predictions obtainable with the selected class. We shall choose the model class as the gaussian ARMA class, which means that we shall have to know how prediction is done optimally for such processes. The Kalman theory in principle is applicable, but the solution it provides involves parameters that cannot be estimated from the observations. For this reason we shall use another approach, Rissanen (1967), and we give the relevant recurrence equations below.

Consider a process generated by the recursion

$$y(t) + a_1 y(t-1) + \cdots + a_p y(t-p) = e(t) + c_1 e(t-1) + \cdots + c_q e(t-q), \tag{3.1}$$

for $t \geq p$, where e is an orthogonal zero-mean process with variance $E(e(t)^2) = \sigma^2$. Letting $u(t)$ for $t \geq p$ stand for the MA process

$$u(t) = e(t) + c_1 e(t-1) + \cdots + c_q e(t-q), \tag{3.2}$$

we see that the covariance $E(u(t)u(s)) = r(t,s),\ t,s \geq p$, satisfies the crucial "bandedness" property

$$r(t,s) = 0, \text{ for } |t-s| > q. \tag{3.3}$$

We let the initial variables be specified by the covariances as follows

$$\begin{aligned} E(u(t)y(s)) &= r(t,s) = 0 \text{ if } t-s > q \\ E(y(t)y(s)) &= r(t,s), \quad t,s \leq p. \end{aligned} \tag{3.4}$$

The problem is to find the orthogonal projection of $y(t)$ on $Y_{0,t-1}$, the subspace spanned by the observations up to $t-1$, written as $\hat{y}(t \mid t-1)$. The task is simple if we find a representation of the process u as follows:

$$u(t) = \varepsilon(t) + c_1(t)\varepsilon(t-1) + \cdots + c_q(t)\varepsilon(t-q)\,,\ t \geq q, \tag{3.5}$$

where $\varepsilon(t)$ is an uncorrelated (but not of unit variance) process; the variables for non-positive indices are zero. The coefficients are found by the Cholesky factorization of the covariance matrix $R = \{r(i,j)\}$ as $R = B'B$, where B is upper triangular,

$$B = \begin{bmatrix} b(0,0) & b(0,1) & . & . & . & b(0,n) \\ 0 & b(1,1) & . & . & . & . \\ . & 0 & . & . & . & . \\ . & . & . & . & . & . \\ . & . & . & . & . & b(n-1,n) \\ 0 & 0 & . & . & . & b(n,n) \end{bmatrix}$$

Specifically, the factors are defined by the following recursions, which also are seen to result from the Gram-Smith orthogonalization procedure,

$$\begin{aligned} b(t-i,t) &= [r(t-i,t) - \sum_{j=i+1}^{q} b(t-j,t)b(t-j,t-i)]b^{-1}(t-i,t-i),\ 1 \leq i \leq q \\ b(t,t) &= +\,[r(t,t) - b^2(t-q,t) - \cdots - b(t-1,t)^2]^{1/2},\ t > 0 \\ b(0,0) &= +\sqrt{r(0,0)}\,,\ b(t-i,t) = 0,\ i > t. \end{aligned} \tag{3.6}$$

We then have

$$c_i(t) = b(t-i,t)b^{-1}(t-i,t-i),\ i = 1, \ldots ,q. \tag{3.7}$$

Since the ε – and the y – processes span one and the same linear space for all n, we easily get the desired recursive equations for the optimal predictor

$$\hat{y}(t\,|\,t-1) + \sum_{i=1}^{q} c_i(t)\hat{y}(t-i\,|\,t-i-1) = \sum_{i=1}^{k} d_i(t)y(t-i), \tag{3.8}$$

where $d_i(t) = c_i(t) - a_i$, $i = 1, \ldots, k$ for $k = \max\{p,q\}$, and the coefficients with undefined index values are zero.

One can show that if the polynomial defined by the coefficients c_i has its roots strictly outside the unit circle, then $c_i(t) \to c_i$. The limiting predictor, then, agrees with the stationary optimal predictor

$$\hat{y}(t\,|\,t-1) + \sum_{i=1}^{q} c_i\hat{y}(t-i\,|\,t-i-1) = \sum_{i=1}^{k}(c_i - a_i)y(t-i) \tag{3.9}$$

We now return to the main problem of how to do the prediction when the coefficients and the two order numbers p and q are not known. We apply the predictive LS principle and proceed as follows. For each pair (p,q) and each t we solve the following ordinary least squares problem

$$\min_{\theta} \sum_{i=1}^{t} \varepsilon^2(i), \tag{3.10}$$

where θ denotes the vector of the coefficients $\alpha = (a_1, \ldots, a_p, c_1, \ldots, c_q)$ together with the $p(p+1)/2 + q(q+1)/2$ initial elements $r(i,j)$ in (3.6), defining a vector β, and $\varepsilon(i)$ may be solved recursively from (3.1), (3.2), and (3.5). Let the minimizing parameters be $\hat{\theta}(t) = (\hat{\alpha}(t), \hat{\beta}(t))$. With these we now extend the Cholesky factorization one more step; i.e., we compute the coefficients (3.6) for $t+1$, and with (3.7) we calculate the new prediction $\hat{y}(t+1\,|\,t)$ from the formula (3.8), which clearly depends only on the past data and the pair (p,q), because the calculation of $\hat{\theta}(t)$ is done by the fixed ordinary least squares algorithm. This gives the prediction error $\bar{\varepsilon}(t+1\,|\,p,q) = y(t+1) - \hat{y}(t+1\,|\,t)$. As the final step we find the best pair $(\hat{p}(n), \hat{q}(n))$ which solves the optimization problem

$$I_P(y) = \min_{p,q} \sum_{t=0}^{n-1} \bar{\varepsilon}^2(t+1 \mid p,q). \tag{3.11}$$

It can be shown that asymptotically

$$\frac{1}{n} I_P(y) \cong \sigma^2 (1 + \frac{p+q}{n} \ln n), \tag{3.12}$$

where

$$\sigma^2 = \frac{1}{n} \sum_{t=1}^{n} \varepsilon^2(t). \tag{3.13}$$

Remarks.

In the above described procedure we did not pay any attention to the amount of computations needed. Rather, our aim was to do the prediction as well as we know how, provided, though, that there is no prior knowledge about the parameter values. Clearly, when calculating $\hat{\theta}(t+1)$ by a suitable hill climbing routine, we should use $\hat{\theta}(t)$ as the initial estimate. It is also possible to calculate the Cholesky factorization by an order of magnitude faster algorithm, Rissanen (1973), in case the covariance matrix $R(t)$ is a Toeplitz matrix; i.e., if the process u is stationary, and if we set the initial conditions to zero. Alternatively, it is enough to have the initial conditions such that the $y-$ process is stationary. The resulting fast predictor recursions have been described by Lindquist (1974), Kailath, Morf, and Sidhu (1974), and by Rissanen (1975), after this author lectured the topic at Stanford University during 1971-1972. Much earlier, the impulse response of a stationary predictor was found with a fast algorithm by Levinson, but that algorithm required an ever growing memory.

The entire Cholesky factorization can be avoided if we ignore the influence of the initial conditions and simply replace the representation (3.5) by (3.2). The only problem remaining then is to compute the sequence of estimates $\hat{\alpha}(t)$, $t = 0, \ldots, n-1$ for different values of p and q. In the case of AR processes such calculations can be done recursively by the so-called ladder forms; see Wax (1985).

As a final remark, the difference between the complexity and the sum of the squared residuals (3.13) was observed in Bittanti (1983), where it was wondered whether the relationship between the two could be clarified. Well, (3.12) does it in a most decisive manner.

We computed in Rissanen (1984c) a small simulation to test the predictive least square principle for estimating an ARMA order. We used the stationary equations which do not require the Cholesky factorization. The data were generated with an ARMA(1,1) system with parameters $a = .5$ and $c = -.3$, where $e(t)$ was a computer generated zero mean unit variance independent gaussian sequence. We fitted models of type ARMA(p,q) with $(p,q) = (1,0)$, (2,0), (1,1), and (2,2). The following table gives the sum in (3.12), calculated for various values of p,q and divided by n, along a single sample of size 600.

(p,q)	n = 50	n = 100	n = 200	n = 300	n = 600
(1,0)	1.336	1.276	1.101	1.107	1.015
(2,0)	1.629	1.385	1.156	1.120	-
(1,1)	1.505	1.307	1.117	1.091	0.996
(2,2)	1.925	1.520	1.221	1.159	-

Table 1. Simulations of ARMA processes

We see that the models (2,0) and (2,2) give uniformly worse values than the two best models (1,0) and (1,1) in the table for all sample sizes (we did not calculate the last entry for them, which surely would have been worse, too). In the last model, in particular, the two extra parameters penalize heavily the prediction errors. For sample sizes up to 200 the simpler model (1,0) performs best, but eventually the model with the right numbers of parameters (1,1) is the winner. This makes sense in that there is no predictive benefit in estimating the second less significant parameter until there is enough data, even if we knew that such a parameter existed; the data are the ultimate arbiter in deciding what is optimal and what is not.

We then wanted to study how initial estimates of the parameters might be taken advantage of to improve the parameter estimates and the predictions. After all, in our opinion, the most natural and easy way to incorporate initial knowledge is directly in terms of the estimate of the parameters, including their numbers. Indeed, the parameters usually represent constants, and any Bayesian type of prior distribution for them is both awkward to justify and just about impossible to estimate in a meaningful way. The traditional Bayesian formalism does not permit a representations of initial knowledge in terms of a parameter value, because the so defined singular distribution cannot be altered by the data. However, our formalism does it easily. In fact, let $\hat{\theta}(0)$ denote the initial estimate with $p+q$ components. Then with $\hat{\theta}(t)$ denoting the predictive LS estimate from the first t observations with initial knowledge ignored, as described above, define a new estimate as a linear combination of the two

$$\tilde{\theta}(t) = \alpha_t \hat{\theta}(0) + (1-\alpha_t)\hat{\theta}(t). \tag{3.14}$$

The coefficient is defined as follows

$$\alpha_t = \frac{1}{1+2^{L_0(p,q,t)-L(p,q,t)}}, \tag{3.15}$$

where $L(p,q,t)$ denotes the accumulated prediction errors, (3.11) before the minimization, and $L_0(p,q,t)$ is the same when the parameter is the initial estimate $\hat{\theta}(0)$. Because this parameter is the same throughout the data, $L_0(p,q,t)$ coincides with the usual non-predictive sum of the squared deviations. We see that a good initial estimate tends to make the corresponding code length shorter than the length $L(p,q,t)$ for small values of t, because initially the estimate $\hat{\theta}(t)$ tends to be poor due to a small sample size. This causes α_t to be near one, and the effective estimate $\tilde{\theta}(t)$ is close to the initial estimate. However, eventually α_t gets small, unless the initial estimate is perfect, and the effective estimate tends to the steadily improving estimate $\hat{\theta}(t)$.

To test the feasibility of this scheme we generated a data sequence of length 100 with the ARMA(1,1) model defined by the two parameters $a = 0.7$, $c = -0.1$, and with a unit variance zero mean gaussian independent input sequence. We set the initial estimate $\theta(0) = (0.7, -0.1)$ of the parameters at the "true" values. We wish to compare the convergence of the parameter estimate $\tilde{\theta}(t) = (\tilde{a}, \tilde{c})$, given by such a perfect initial knowledge, with that of the least squares estimates $\hat{\theta}(t) = (\hat{a}, \hat{c})$. In this experiment we, then, kept the number of parameters at the correct value. These estimates along with the two sums of the squared predictions, corresponding to the two estimators $\tilde{\theta}$ and $\hat{\theta}$, respectively, were computed along the 100 sample points, and the results are in the following table.

	with prior estimates			without prior estimates		
time t	a	c	L	a	c	L
10	0.15	0.01	5.0	0.04	0.03	4.99
20	0.47	-0.23	17.1	0.16	-0.41	20.0
30	0.62	-0.17	28.7	0.24	-0.50	33.7
40	0.69	-0.11	36.0	0.43	-0.28	42.9
100	0.69	-0.11	98.9	0.50	-0.33	108.2

Table 2. Effect of prior estimates

We see that indeed good initial estimates improve both the convergence and the prediction error.

4. Vector Time Series Models

As quite well known, the class of multi-input/output linear dynamic systems, even of a fixed dimensionality, is topologically a lot more complex space than in the case when either the input or the output is scalar. Hence, when we search for the stochastic complexity of an observed vector time series, relative to the class of such models, we may find a model which with relatively few parameters will capture the essence in the data. In the older statistical literature the only models that were fitted to a series with, say, p components, had the maximal dimensionality, a multiple of p. This was justified on the grounds that since the estimated Hankel-matrix, or its equivalent, has the maximal rank, there is no point in fitting other models. Such an argument indicates a gross misunderstanding of modeling, and, in fact, an equivalent argument would dismiss fitting dynamic systems to scalar sequences as well; after all, no observed sequence is generated by any dynamic system.

Since the theory of multivariable linear dynamic systems is by now well known, and, in fact, it may even be covered in some of the other chapters of this book, we do not need to describe it here in any detail. Instead, we just summarize the relevant facts. The set of all linear dynamic systems with, say, p inputs and equally many outputs, is in one-to-one correspondence with the set of all Hankel matrices of $p \times p$ blocks with finite rank, say n. Given such a system, its matrix input/output impulse response defines a Hankel matrix of $p \times p$ blocks and rank n. Conversely, any of the usual realization algorithms defines a p-input/output system of degree n of such a Hankel matrix.

The set of all finite rank Hankel matrices of $p \times p$ blocks clearly admits a partition into equivalence classes by the rank n. How can each such class be parameterized? Unlike in the case with $p = 1$, an equivalence class corresponding to a rank n, and hence the set of all linear systems of order n with p inputs and outputs, cannot be parameterized with a single coordinate system, and the set is not a linear space. This important observation is at the root of the modern theory of linear dynamic systems, and it also affects in a profound manner the way such models ought to be fitted to the observed time series. Consider, for example, the set of all Hankel matrices with $p = 2$ and $n = 3$. If we further assume the first two rows to be linearly independent, as we should to avoid pathology, then the Hankel property implies that either the third or the fourth row must be the last remaining row that together with the first two forms a 3-element basis for the span of all the rows in the matrix. Again the Hankel property implies that these three basis rows are defined just as soon as we specify the two first elements in each, hence, six alltogether. In the former case, where the basis consists of the first three rows, the fourth and the fifth row are linear combinations of the basis elements and, hence, to specify them we need six coefficients. All the other rows in the Hankel matrix are now just shifts and truncations of these and the basis rows. Similarly, $2np = 12$ parameters are needed to specify all the Hankel matrices, where the first, second, and the fourth rows form a basis.

Consider now the set of matrices where the fourth row is a basis element. Then the third row per force is linearly dependent on the first, second, and the fourth. Consider the further subset where the third row in fact is linearly dependent on the first two. Evidently no such matrix and the corresponding linear system could be expressed in terms of the parameters defined by the basis consisting of the first, second, and the third row. From this we conclude that in order to parameterize the set of all systems of degree 3 having two inputs and outputs, we need two distinct coordinate systems.

In general, then, the set of all linear systems of degree n having p inputs and outputs, may be partitioned into finitely many equivalence classes, each class corresponding to the so-called lexicographic basis defined by each matrix as follows: Each of the first p rows is included in the basis, and the next basis element is the first row which is not in the linear span of those above it, and so on. Consider the ith row, $i \leq p$, and let n_i be the least multiple of p such that the row with index $i + pn_i$ is not in the basis. The set of the p positive numbers, called observability indices, $s = (n_1, \ldots, n_p)$, defines a *structure*. Just as in the example above, $2np$ coefficients in each coordinate system define uniquely a Hankel matrix and, hence, a linear dynamic system, and we see that the set of all systems having the same structure defines a $2np$-dimensional Euclidean space. Although each Hankel matrix defines a unique lexicographic basis, it is still possible that some (and, in fact, most) matrices can be and get defined in several distinct bases. In the example above, a Hankel matrix may well have both the first, second, and the third rows as well as the first, second, and the fourth rows as linearly independent. However, no single coordinate system suffices to describe them all. In technical terms, the set of all systems of the same degree is an analytic manifold, which was recognized by Clark (1976), whose work culminates the efforts by Kalman (1974), Glover and Willems (1974), and Rissanen and Ljung (1975). We mention to this end another closely related the so-called canonical or minimal parameterization of systems. This is obtained by constructing a partition from the over-lapping subsets of systems, defined by the coordinate systems in the manifold. The advantage of the first representation is precisely the fact that the coordinate systems cover overlapping subsets, which means that there are smooth coordinate transformations between them.

The state space representation of the multi-input/output linear systems is somewhat simpler than the ARMA representation for the reason that the invariants; i.e., the coordinates, appear directly as elements in the system matrices; in the ARMA representation the matrix elements are functions of the invariants. What we need is the equations for the predictions, which we for simplicity take to be time invariant, which means that we need not solve the Riccati equations or, equivalently, find the Cholesky factors recursively; otherwise, we would have to proceed as in the previous section. The predictor equations, then, are as follows:

$$\begin{aligned} \hat{x}(t+1|t) &= F\hat{x}(t|t-1) + Gu(t) + K[y(t) - H\hat{x}(t|t-1)] \\ \hat{y}(t|t-1) &= H\hat{x}(t|t-1), \end{aligned} \tag{4.1}$$

where $\hat{y}(t|t-1)$ is the prediction of the observed p-component output sequence, u is the possibly present observed r-component input sequence, and $\hat{x}(t|t-1)$ is the prediction of the intermediate n-component state sequence, which we, otherwise, have no interest in. Take initially $\hat{x}(1|0) = 0$. The observed input sequence is quite irrelevant; its number of components, whether single or multiple or none at all, has no important effect on the theory. It is the number of outputs p that we now require to be greater than unity. We may take all the elements in the two matrices G and K as free. The matrix F has np free parameters in the manifold representation, and possibly fewer in its minimal representation. Their locations depend on the choice of the coordinate system, which also determine the location of the 0's and 1's in H, which are its only elements. Hence, there are k free parameters in the model (4.1), which we arrange into a vector θ. In the case with the manifold representation, $k = n(2p + r)$ in each of the structures corresponding to the dimensionality n, while in the minimal realization k depends on the structure; however, $k \leq n(2p + r)$.

If we agree to measure the prediction errors by squares, we get the so-called Gauss - Markov class of models. Relative to this class the non-predictive stochastic complexity of the data $(y|u) = (y(1)|u(1)), \ldots ,(y(N)|u(N))$ is given by

$$I_{NP}(y|u) = \min_{s,\theta} \{\frac{1}{2} \log \det R(\theta) + \frac{k}{2} \log N + k \log \|\theta\|_{M(\theta)}\}, \tag{4.2}$$

where

$$R(\theta) = \frac{1}{N}\sum_{t=1}^{N}(y(t) - \hat{y}_\theta(t|t-1))(y(t) - \hat{y}_\theta(t|t-1))' \tag{4.3}$$

and $M(\theta)$ denotes the Hessian defined by the double derivatives of log det $R(\theta)$, evaluated at θ. It was shown in Rissanen (1983b) that if, indeed, the data were generated by some such system in a structure s, then the estimated structure $\hat{s}(N)$ and the associated parameters $\hat{\theta}(N)$, will approach the corresponding generating parameters. In particular, the last term in the complexity forces the structure estimates to converge. The last two terms in the stochastic complexity (4.2) represent the optimal model complexity as measured in terms of the number of binary digits required to encode the parameters. As the number of data points grows, the second term becomes dominantly greater of the two, and we see that the optimal model complexity grows proportional to $\frac{\hat{k}(N)}{2}\log N$. This is because one must express the parameters collectively with increasing precision as the sample size grows, which makes eminent sense.

We next describe the predictive stochastic complexity. It is given by

$$I_P(y\,|\,u) = \min_s \frac{1}{2}\log\det\frac{1}{N}\sum_{t=1}^{N}(y(t)-\hat{y}(t\,|\,t-1))(y(t)-\hat{y}(t\,|\,t-1))', \tag{4.4}$$

where $\hat{y}(t\,|\,t-1)$ denotes the prediction obtained as follow: The maximum likelihood estimate $\hat{\theta}(t-1)$ is determined from the data up to time $t-1$. With this parameter the predictions $\hat{y}(i\,|\,i-1)$ are computed from (4.1) up to time $i=t$, which gives the prediction needed in (4.4) at this time instant. The whole process is to be repeated for the next time instant, which requires a lot of computations. The bulk of these goes to the evaluations of the estimates $\hat{\theta}(t)$, which must be done for each t. It is clear that $\hat{\theta}(t)$ provides an excellent starting point for getting $\hat{\theta}(t+1)$, but nevertheless an order of t computations are needed to calculate $\hat{y}(t\,|\,t-1)$, so that the entire string takes an order of N^2 operations.

The stochastic complexity (4.4) involves the minimization with respect to the structure index s. If we represent the models in the manifold, then the predictors computed in two equivalent coordinate systems are identical, to within the numerical precision. Therefore, equivalent coordinate systems also produce the same value for the stochastic complexity, and cannot be distinguished. In the minimal representations there are no equivalent coordinate systems and the predictive stochastic complexity is given by a unique structure index. These facts can be turned to an advantage in that for each order n we evaluate the accumulated prediction errors in (4.4) (before minimization) by calculating the required least squares estimates $\hat{\theta}(t)$ in a good structure $\hat{s}(n)$, determined suitably, say, with the procedure in van Oberbeek and Ljung (1982). This is important to ensure that the estimates can be determined with sufficient precision. Then the minimization required in (4.4) may be done with respect to n, just as in the single input/output case, with $\hat{n}$ as the result. Finally, we may also find the minimizing structure $\hat{s}(\hat{n}) = (\hat{n}_1, ..., \hat{n}_p)$ by letting the models be in their minimal representations.

Simulations.

These simulations were done by V. Wertz, see Rissanen and Wertz (1985). The data were generated by the system, Example 5.1 in van Oberbeek and Ljung (1982),

$$\begin{aligned} x(t+1) &= F x(t) + G u(t) + K e(t) \\ y(t) &= H x(t) + e(t), \end{aligned} \tag{4.5}$$

which in the manifold representation has the matrices

$$F = \begin{bmatrix} 0 & 1 & 0 \\ .25 & 0 & .25 \\ 1 & 0 & 0 \end{bmatrix}, \quad G = \begin{bmatrix} 0 \\ 1 \\ 0 \end{bmatrix}$$

and

$$K = \begin{bmatrix} .000547 & .063 \\ .119 & .157 \\ .674 & .0000666 \end{bmatrix} \quad H = \begin{bmatrix} 1 & 0 & 0 \\ 0 & 0 & 1 \end{bmatrix}.$$

This system has the structure $s = (2,1)$ corresponding to the basis defined by the first three rows in the Hankel matrix. The 2-component input $e(t)$ is a sample from a 2-variate zero mean gaussian process with the covariance matrix given by $.2 \times I$, where I denotes the 2×2 identity matrix. The scalar input $u(t)$ is an observed more or less random signal, taking values +1 or -1, which was added for good measure. The sample size is N = 500; i.e., t = 1, 2,..., 500.

Three different models with reasonable structures $s_1 = (1,1)$, $s_2 = (2,1)$, and $s_3 = (2,2)$ were fitted. The corresponding values of the minimized accumulated prediction errors are .3031, .0649, and .0671, respectively. We see that the second structure, which is the same as the structure of the data generating system, gives the predictive complexity. The first being under-parameterized is clearly the worst, while the last, having only one excess parameter, suffers only slightly for this. We give the optimal matrices for the winning model

$$F = \begin{bmatrix} 0 & 1 & 0 \\ .25 & -.02 & .26 \\ 1 & -.02 & .00 \end{bmatrix}, \quad G = \begin{bmatrix} .008 \\ .032 \\ .0490 \end{bmatrix}$$

and

$$K = \begin{bmatrix} -.11 & .085 \\ .73 & -.023 \\ -.11 & .083 \end{bmatrix}, \quad H = \begin{bmatrix} 1 & 0 & 0 \\ 0 & 0 & 1 \end{bmatrix}.$$

References

Abramson, N. (1968), Information Theory and Coding. McGraw-Hill, New York.

Akaike, H. (1974), "A New Look at the Statistical Model Identification", IEEE Trans. AC-19, 716-723.

Akaike, H. (1975), "Markovian Representation of Stochastic Processes by Canonical Variables", SIAM J. Control, 13.

Bittanti, S. (1983), "Is the Prediction Error of a Regression Model White?", J. Franklin Inst. Vol. 315, No. 4, 239-246.

Chaitin, G.J. (1975), "A Theory of Program Size Formally Identical to Information Theory" J.ACM, 22, 329-340.

Clark, J.M.C. "The Consistent Selection of Local Coordinates in Linear Systems Identification", JACC, Purdue University, Lafayette, Indiana, pp. 576-580, July 1976.

Dawid, A.P. (1984), "Present Position and Potential Developments: Some Personal Views, Statistical Theory, The Prequential Approach", J. Royal Stat. Soc. Series A, Vol. 147, Part 2, 278-292.

Geisser, S. and Eddy, W. (1979), "A Predictive Approach to Model Selection", J. American Stat. Ass., Vol. 74, Nr. 365, 153-160.

Glover, K. and Willems, J.C. (1974), "Parameterizations of Linear Dynamical Systems: Canonical Forms and Identifiability", IEEE Trans. AC-19, no. 6.

Hannan, E.J. (1980), "The Estimation of the Order of an ARMA Process", Ann. Stat. 8, No. 5, 1071-1081.

Hjorth, U. (1982), "Model Selection and Forward Validation", Scand. J. Stat. 9, 95-105.

Kailath, T., Morf, M., Sidhu, G.S. (1974), "Some New Algorithms for Recursive Estimation in Constant Discrete-Time Linear Systems", IEEE Tr. Automatic Control, Vol. AC-19, 315-323.

Kalman, R.E. (1974), "Algebraic Geometric Description of the Class of Linear Systems of Constant Dimension", 8'th Ann. Princeton Conf. on Inf. Sciences and Systems, Princeton, New Jersey.

Kolmogorov, A.N. (1965), "Three Approaches to the Quantitative Definition of Information", Problems of Information Transmission 1, 4-7.

Lindquist, A. (1974), "A New Algorithm for Optimal Filtering of Discrete-Time Stationary Processes", SIAM J. Control 4, 736-747.

Ljung, L. and Rissanen, J. (1976) "On Canonical Forms, Parameter Identifiability and the Concept of Complexity", IFAC Symp. on Identification, Tbilisi, USSR.

Luenberger, D.G. (1974), "Canonical Forms for Linear Multivariable Systems", IEEE Trans. AC-12, 290-293.

Popov, V.M. (1972), "Invariant Description of Linear, Time-Invariant Controllable Systems", SIAM J. Control, 10, 254-264.

Rissanen, J. (1967), "An algebraic approach to the problems of linear prediction and identification", IBM Res. Rep. RJ 468, Oct. 23.

Rissanen, J. (1973), "Algorithms for Triangular Decomposition of Block Hankel and Toeplitz Matrices with Application to Factoring Positive Matrix Polynomials", Mathematics of Computation, Vol. 27, 147- 154.

Rissanen, J. (1974), "Basis of Invariants and Canonical Forms for Linear Dynamic Systems", Automatica, Vol. 10, pp.175-182.

Rissanen, J. (1975), "Canonical Markovian Representations and Linear Prediction", Proc. of the 6'th IFAC Symposium, Part 1, Paper 29.3, 1-9.

Rissanen, J. (1978), "Modeling by shortest data description", Automatica, Vol. 14, pp. 465-471.

Rissanen, J. (1983a), "A Universal Prior for Integers and Estimation by Minimum Description Length", Ann. of Statistics, Vol. 11, No. 2, 416-431.

Rissanen, J. (1983b), "Estimation of Structure by Minimum Description Length", Circuits, Systems, and Signal Processing, special issue on Rational Approximations, Vol. 1, Nr. 3-4, 395-406.

Rissanen, J. (1984a), "Universal Coding, Information, Prediction, and Estimation", IEEE Trans. Inf. Theory, Vol. IT-30, Nr. 4, 629-636.

Rissanen, J. (1984b), "Stochastic Complexity and Modeling", (to appear in Ann. of Statistics).

Rissanen, J. (1984c), "Order Estimation by Accumulated Prediction Errors", Esseys in Time Series and Allied Processes (eds. J. Gani, M.B. Priestley).

Rissanen, J. (1985a), "Minimum Description Length Principle", Encyclopedia of Statistical Sciences, Vol. V, (S. Kotz & N. L. Johnson eds.), pp. 523-527. John Wiley and Sons, New York.

Rissanen, J. (1985b), "A Predictive Least Squares Principle", (to appear).

Rissanen, J. and Ljung, L. (1975), "Estimation of Optimum Structures and Parameters for Linear Systems", Proc. CNR. CISM Symp. on Algebraic System Theory, Udine, Math. System Theory 131, Springer-Verlag, pp. 76-91.

Rissanen, J. and Wertz, V. (1985), "Structure Estimation by Accumulated Prediction Error Criterion", Eighth IFAC Symposium on Identification and System Parameter Estimation, York, England.

Schwarz, G. (1978), "Estimating the Dimension of a Model", Ann. Statist. 6, 461-464.

Shibata, R. (1976), "Selection of the Order of an Autoregressive Model by Akaike's Information Criterion", Biometrica, 63, 1, 117-126.

Solomonoff, R.J. (1964), "A Formal Theory of Inductive Inference". Part I, Information and Control 7, 1-22; Part II, Information and Control 7, 224-254.

Stone, M. (1977), "An Asymptotic Equivalence of Choice of Model by Cross-Validation and Akaike's Criterion", J. Royal Stat. Soc., Ser. B, 39, 44-47.

van Overbeek, A.J.M. and Ljung, L. (1982), "On Line Structure Selection for Multivariable State Space Models", Automatica, vol. 18. no. 5, 529- 543.

Wax, M. (1985), "Order Selection for AR Models by Predictive Least Squares", (to appear)

Wertz, V. (1982), Structure Selection for the Identification of Multivariate Processes, Dr. Sci. Appl. thesis, Universite Catholique de Louvain, Louvain-La-Neuve.

Wertz, V., Gevers, M., Hannan, E.J. (1982), "The Determination of Optimum Structures for the State Space Representation of Multivariable Stochastic Processes", IEEE Trans. Autom. Control., Vol. AC-27, No.6, 1200 - 1211.

Chapter 5

Deterministic and Stochastic Linear Periodic Systems

Sergio Bittanti

1. Introduction

Linear periodic systems are linear systems described by differential or difference equations with periodically time-varying coefficients. Deterministic and stochastic periodic systems are useful to model natural and artificial phenomena of periodic type. As such, they are of great interest in various fields of application, such as Chemical and Aerospace Engineering, Signal Modelling and Processing. Moreover, periodic systems play a key role in optimal periodic control theory. This theory stands on the observation that there exist several systems of practical significance for which the best operation is a periodic one. The implementation of such a periodic action in presence of stochastic disturbances calls for a detailed analysis of the periodic control problem of stochastic periodic systems. Without any claim of completeness, the following general references relative to the area of Periodic Systems and Control (PSYCO) (theory and applications)are mentioned: (Bailey, 1973), (Bekir and Bucy, 1976) (Berstein and Gilbert, 1980), (Bittanti, Fronza and Guardabassi, 1973), (Colonius,1985a, b, c), (DaPrato, 1984), (Dorato and Levis, 1971), (Dorato and Knudsen, 1979), (Gilbert, 1977), (Gilbert and Lyons, 1981), (Guardabassi, 1971 and 1976), (Hernandez and Jodar, 1985), (Horn and Lin, 1967), (Horn and Bailey, 1968), (Houlinhan, Cliff and Kelley,

1982), (Khargonekar,Poolla and Tannenbaum, 1985), (Kern, 1980), (Khandelwal, Sharma and Ray, 1979), (Kono, 1980), (Maffezzoni, 1974), (Marcus, 1973), (Matsubara, Nishimura, Watanabe and Onogi, 1981), (Matsubara and Onogi, 1978), (Meyer and Burrus, 1976), (Nistri,1983), (Noldus, 1975), (Onogi and Matsubara, 1980), (Schädlich, Hoffmann and Hofmann, 1983), (Shayman, 1985), (Sincic and Bailey, 1978), (Speyer, 1973 and 1976), (Speyer and Evans, 1984), (Watanabe, Nishimura and Matsubara, 1984), (Watanabe, Nishimura and Matsubara, 1976), (Watanabe, Onogi and Matsubara, 1981), (Watanabe, Kurimoto and Matsubara, 1984). Further references will be quoted in the sections below.

Obviously, the linear time-invariant systems belong to the class of linear periodic systems. However, the extension of the properties of time-invariant systems to the periodic case is far from straightforward, as witnessed by the very development of the theory of PSYCO. Indeed, many peculiar and challenging problems are encountered along this route. Only in the last few years, a number of these problems have been solved and some open questions have been clarified.

In particular, a somewhat detailed understanding of the structural properties of periodic systems has been achieved. This paper is intended to provide a first general picture of such a subject by surveying the appropriate literature, which covers the two last decades. The use of these properties in the study of some basic questions relative to stochastic linear periodic systems is also discussed.

The paper is expository in nature and is organized as follows. The structural properties (reachability, controllability, and so on) of continuous-time and discrete-time systems are dealt with in Sections 2 and 3 respectively. Precisely, five well known properties valid in the time-invariant case are first recalled. Then, their generalization to periodic systems is discussed. The Kalman canonical decomposition in terms of these properties is the subject of Section 4. Then, Section 5 deals with the extended structural properties, i.e. stabilizability and detectability. **Finally,** Section 6 is devoted to the analysis of stochastic systems. Precisely, the existence of a cyclostationary stochastic solution is investigated.

2. Structural Properties of Continous-time Periodic Systems

2.1 Continous-time Linear Periodic Systems

Consider the system described by the differential equation:

$$\dot{x}(t) = A(t)\ x(t) + B(t)\ u(t) \qquad (1.a)$$

where $A : R \to R^{nxm}$ and $B : R \to R^{nxn}$ are continuous real and T-periodic.The period T is the smallest value for which

$$A(t+T) = A(t) \quad ; \quad B(t+T) = B(t) \quad , \forall t. \qquad (1.b)$$

The system transition matrix $\Phi(t,\tau)$, i.e. the solution of

$$\frac{d}{dt}\ \Phi(t,\tau) = A(t)\ \Phi(t,\tau) \quad , \quad \Phi(\tau,\tau) = I, \quad t>\tau,$$

is such that

$$\Phi(t+T, \tau+T) = \Phi(t,\tau). \qquad (2)$$

By the Floquet theory, see e.g. Halanay (1966), Chen (1970) and Yakubovich and Starzhinskii (1975), $\Phi(t,0)$ can be expressed as follows:

$$\Phi(t,0) = \Psi(t)\, e^{Rt}$$

where

$$\Psi(t+T) = \Psi(t), \quad \forall t; \quad \Psi(0) = I.$$

The matrix $\Phi(t+T,t)$ is named *monodromy matrix* at time t. In particular $\Phi(T,0)=e^{RT}$ is the monodromy matrix at t=0, or simply *the* monodromy matrix. Since $\Phi(t+T,t)= \Phi(t,\tau)\, \Phi(\tau+T,\tau)\, \Phi(t,\tau)^{-1}$, the eigenvalues of $\Phi(t+T,t)$ are independent of t. They are called the *characteristic multipliers*. By Jacobi's Theorem, the product of these eigenvalues must be positive. In fact :

$$\det(e^{RT}) = e^{(\mathrm{tr}\, R)T} \quad .$$

The eigenvalues of R are named *characteristic exponents*.

System (1) is asymptotically stable if and only if the characteristic multipliers lie within the open unit disk, which is equivalent to requiring that the characteristic exponents belong to the open left plane.

The spectrum of e^{RT} will be denoted by Σ, while Σ_1 is used to denote the set of the "unstable" eigenvalues of the same matrix (namely the eigenvalues not belonging to the open unit disk).

2.2 Structural properties

For the sake of completeness, the classical definitions of reachability and controllability are given below.

Definition 1

1.1 The state $x \in R^n$ is reachable over $(\tau,t), \tau < t$, [controllable over $(t,\tau), \tau > t$] if there exists an input function for (1.a) which carries the event $(\tau,0)$ into (t,x) [(t,x) into $(\tau,0)$].

1.2 $X_r(\tau,t)$ [$X_c(t,\tau)$] denotes the set of the reachable [controllable] states over (τ,t) [(t,τ)].

1.3 System (1.a) (or, equivalently, the pair (A,B)) is reachable [controllable] over (τ,t) [(t,τ)] if $X_r(\tau,t) = R^n$ [$X_c(t,\tau) = R^n$].

1.4 The state $x \in R^n$ is reachable at t [controllable at t] if there exists a time point $\tau, \tau < t$ [$\tau > t$], such that x is reachable over (τ,t) [controllable over (t,τ)].

1.5 $X_r(t)$ [$X_c(t)$] denotes the set of the reachable [controllable] states at t.

1.6 System (1.a) (or, equivalently the pair (A,B)) is reachable [controllable] at time t if $X_r(t) = R^n$ [$X_c(t) = R^n$].

1.7 System (1) (or, equivalently, the pair (A,B)) is reachable [controllable] if $X_r(t) = R^n, \forall t$ [$X_c(t) = R^n, \forall t$].

2.3 Grammian matrices

The following nxn matrices are named *reachability* and *controllability Grammian* matrices respectively.

$$W_r(\tau,t) = \int_{\tau}^{t} \Phi(t,\sigma)\ B(\sigma)\ B(\sigma)'\ \Phi(t,\sigma)'\ d\sigma\ ,\ t > \tau \tag{3.a}$$

$$W_c(t,\tau) = \int_{t}^{\tau} \Phi(t,\sigma)\ B(\sigma)\ B(\sigma)'\ \Phi(t,\sigma)'\ d\sigma\ ,\ \tau > t \tag{3.b}$$

It is well known (Kalman, 1969) that

$$X_r(\tau,t) = R\left[W_r(\tau,t)\right]$$

$$X_c(t,\tau) = R\left[W_c(t,\tau)\right]$$

where $R\left[\cdot\right]$ is the range operator.
In the periodic case, the following recursions can be derived in view of (2):

$$W_r(t-(i+1)T,t) = W_r(t-iT,t) + \left[\Phi(t+T,t)\right]^i W_r(t-T,t) \left[\Phi(t+T,t)'\right]^i \tag{4.a}$$

$$W_c(t,t+(i+1)T) = W_c(t,t+iT) + \left[\Phi(t+T,t)\right]^{-i} W_c(t,t+T) \left[\Phi(t+T,t)'\right]^{-i} \tag{4.b}$$

2.4 Five structural properties of time-invariant systems

The structural properties of linear time-invariant systems have received ample coverage in the literature, see e.g. Kalman (1969),

Chen (1970). Five well known properties are listed below. Although some of them are trivial, it is advisable to list them all, in the proper order, for the subsequent discussion on periodic systems.

A) The reachability and controllability subspaces at time t do coincide:

$$X_r(t) = X_c(t) \quad , \quad \forall t$$

B) The reachability and controllability subspaces are time-invariant:

$$X_r(t) = \text{const.} \quad , \forall t$$

$$X_c(t) = \text{const.} \quad , \forall t.$$

C) If the pair (A,B) is reachable [controllable] at a time point, it is reachable [controllable] at any time point:

$$X_r(t) = R^n \quad \Longrightarrow \quad X_r(t) = R^n \; , \forall t$$

$$X_c(t) = R^n \quad \Longrightarrow \quad X_c(t) = R^n \; , \forall t.$$

D) If a state is reachable at t [controllable at t], then, for any $\varepsilon>0$, it is reachable over $(t-\varepsilon,t)$ [controllable over $(t, t+\varepsilon)$] :

$$X_r(t) = X_r(t-\varepsilon,t)$$

$$X_c(t) = X_c(t, t+\varepsilon).$$

E) The pair (A,B) is reachable if and only if

$$\begin{bmatrix} sI - A & B \end{bmatrix}$$

is full rank on the spectrum of A.

Due to the system time-invariance, Property B is intuitive. C is a trivial consequence of B and is listed here only to stress the difference between the time-invariant and the periodic case. Property D can be rephrased by saying that, in time-invariant systems, reachability and controllability are "asymptotically instantaneous". The reachability and controllability transitions can be made to occur in an interval of time arbitrarily short. This is connected with the absence of any energy constraint on the input function in the classical Definitions 1. Finally, the spectral characterization E is often referred to as the PBH (Popov-Belevitch-Hautus) condition, see Johnson (1966), Popov (1966), Belevitch (1968), Hautus (1969).

2.5 Five structural properties of continuous-time periodic systems

The following basic questions are considered in this section:

. Do properties A-C hold true for periodic systems?
. Can anything be said about the reachability and controllability intervals?
. Does there exist a periodic version of the PBH test?

In the first place, A still holds true, i.e. the reachability and controllability subspaces coincide even in the periodic case: $X_r(t) = X_c(t)$, $\forall t$.

As many results concerning the structural properties of linear periodic systems, this can be proven either via algebraic or geometric methods.

As an example of a typical geometric proof, let us outline the derivation of inclusion $X_c(t) \subseteq X_r(t)$. Let x^* be a state which is controllable at t and denote by N a positive integer such that any state controllable at t can be driven to zero in an interval NT. Due to periodicity, $X_c(t) = X_c(t+NT)$, so that x^* is controllable at t+NT as well.
Let $\bar{x} = \Phi(t,+NT)x^*$. It is apparent that, since x^* is controlable at t+NT, $\bar{x}$ is controllable at t. Therefore there exists an input function which transfers $(t,\bar{x})$ into (t+NT,0). By the same input function, the event (t,0) will then be transferred to $(t+NT,-x^*)$. Therefore x^* is reachable at t+NT, and consequently reachable at t thanks to periodicity. This leads to the conclusion $X_c(t) \subseteq X_r(t)$.

The coincidence of the reachability and controllability subspaces of a periodic system at each time point is an extension of the analogous well known property for time-invariant systems. However, contrary to the time invariant case, the subspaces $X_r(\tau,t)$ and $X_c(\tau,t)$ may not coincide (see Example 1 below).

Property B is obviously false. Instead $X_r(t)$ and $X_c(t)$ are periodically time-varying :

$$X_r(t) = X_r(t+T) \quad , \quad \forall t$$

$$X_c(t) = X_c(t+T) \quad , \quad \forall t.$$

However, by the same arguments used in discrete-time in Bittanti and Bolzern (1984,a), Lemma 3, it can be proven that

$$\dim X_r(t) = \text{const}$$

$$\dim X_c(t) = \text{const}.$$

From this, it follows that property C still holds true: If a periodic system is reachable at a given time point, it is reachable at any time point. Hence it is possible to speak of system reachability and controllability without further specifications.

The attention is now focused on properties D and E, the stories of which are somewhat interwoven with each other. To the best knowledge of the present author, the first statements concerning the length of the reachability and controllability intervals of periodic systems go back to the late sixties. In Brunovsky (1969), the following result is proven by means of algebraic arguments. An alternative proof, of geometric type, can be found in Bittanti and Bolzern (1984,a), Lemma 1.

<u>Theorem 1</u> (Brunovsky, 1969)

If system (1) is controllable, then the controllability transition can be performed in an interval of time of length nT (n is the system order) :

$$X_c(t) = R^n \implies X_c(t) = X_c(t,t+nT). \square$$

In Kalman (1969), a stronger statement is reported without proof.

<u>Proposition</u> (Kalman, 1969)

If system (1) is controllable, then the controllability transition can be performed in an interval of time of length T. □

The question of proving Kalman proposition remained open untill 1975, when, in a lengthy paper on the periodic Riccati equation, Hewer gave a proof of Kalman proposition. Furthermore, he gave a spectral condition of system controllability which generalizes the PBH test to the periodic case.

A slightly different yet equivalent condition is due to Kano and Nishimura (1979), where reference is made to the monodromy matrix $\Phi(T,0) = e^{RT}$ in place of R. The condition, named H-condition, will now be stated in terms of a generic monodromy matrix $\Phi(t+T,t)$. This is especially useful for the extension to discrete-time systems. Recall that $\sum$ is the spectrum of e^{RT}.

<u>H-condition at t</u> (continuous-time)

Given a time point t, the matrix

$$\left[sI - \Phi(t+T,t) \quad W_c(t,t+T)\right]$$

is full rank on $\sum$. □

The first paper where the condition was stated in these terms is probably Bittanti, Bolzern, Colaneri and Guardabassi (1983).

However, the spectral conditions played a key role in the analysis of the periodic Lyapunov and Riccati Equations, Hewer (1969), Kano and Nishimura (1975). The proof given by Hewer of the validity of the H-condition as controllability test was based on Kalman proposition, though. Unfortunately, the proof given in Hewer (1969) of such a proposition was not correct. Even more so: the Kalman proposition itself is not true, as shown by the following counterexample.

Example 1

For a given integer n, let $\lambda_1, \lambda_2, \ldots, \lambda_n$ be n given distinct real numbers. Consider the single-input system:

$$A(t) = \text{diag}\left[\lambda_1, \lambda_2, \ldots, \lambda_n\right],$$

$$B(t) = \begin{cases} \left[e^{-\lambda_1(1-t)} \quad e^{-\lambda_2(1-t)} \quad \ldots \quad e^{-\lambda_n(1-t)}\right]' \sin \pi t, & t \in [0,1] \\ \text{periodic extension of previous}, & t \notin [0,1]. \end{cases}$$

For this system, which is periodic of period T = 1,

$$\Phi(0,\sigma)\, B(\sigma) = (\sin \pi\sigma)\, x_1$$

where

$$x_1 = \left[e^{-\lambda_1} \quad e^{-\lambda_2} \quad \ldots \quad e^{-\lambda_n}\right]'$$

Letting

$$\alpha = \int_0^1 \sin^2 \pi\sigma \, d\sigma$$

it follows from (3.b) that

$$W_c(0,1) = \alpha\, x_1 x_1'.$$

Therefore, dim $X_c(0,1) = 1$, and, assuming $n > 1$, this system is not controllable over $(0,1)$.

For a given positive integer k, $k \leq n$, consider now the time interval $(0,k)$ and let

$$x_i := \left[e^{-i\lambda_1} \quad e^{-i\lambda_2} \ldots e^{-i\lambda_n}\right]', \qquad i = 1, 2, \ldots, k.$$

Then, by applying recursion (4) for any integer $k \leq n$, $W_c(0,k)$ can be given the following expression:

$$W_c(0,k) = \alpha(x_1x_1' + x_2x_2' + \ldots + x_kx_k'), \quad k < n$$

Consequently,

$$X_c(0,k) = \text{span}\left[x_1, x_2, \ldots, x_k\right], \qquad k \leq n.$$

Since $x_1, x_2, \ldots, x_n$ are independent, it follows that

$$\dim X_c(0,k) = k \quad , \quad \forall k \leq n.$$

Therefore the system is controllable, but there exist some states which cannot be driven to zero in an interval of time shorter than (n).

Interestingly enough, it turns out that

$$X_r(0,k) = \text{span}\left[x_0, x_{-1} \ldots, x_{-k+1}\right]$$

This entails that $X_r(0,k)$ and $X_c(0,k)$ may not coincide. □

The validity of the H-condition as a controllability condition must now be proven indipendently of Kalman's conjecture. Notice that, although the reachability and controllability transitions cannot be made to occur in a single period, the H-condition calls for the controllability Grammian over a *single* period.
One such proof, can be found in Bittanti, Bolzern, Colaneri and Guardabassi (1983) and Bittanti, Colaneri and Guardabassi (1984), which contain the following results. In Theorem 2(b), ν is the degree of the minimal polynomial of the monodromy matrix at any time point, $\Phi(t+T,t)$. Note that the minimal polynomial of $\Phi(t+T,t)$ does not depend upon t, see Bittanti, Bolzern, Colaneri and Guardabassi (1983).

Theorem 2

(a) If system (1) is controllable at time t, then the H-condition at t is satisfied.

(b) Suppose that the H-condition at t is satisfied. Then, the system is controllable at t and $X_c(t) = X_c(t,t+\nu T)$. □

In conclusion, the H-condition is in fact a controllability condition. In view of the PBH test, this means that the periodic system (1) is controllable at t if and only if the time-invariant system defined by the pair $(\Phi(t+T,t), W_c(t,t+T))$ is controllable. Since a periodic system is controllable at any time point whenever it is controllable at a given time point, it readily follows from part (b) that, if the H-condition is satisfied at a given time point, then the same condition holds true at any time point. Finally, it is clear that Brunovsky's theorem can be slightly strengthened by claiming that, if system (1) is controllable at t, the controllability transition can be performed in an interval of time of length νT.

Remarks

1. The above is the strongest conclusion on the controllability interval length of a controllable system which can be drawn in terms of matrix A only. This means that, given any periodic matrix A, there exists a periodic matrix B such that, if (A,B) is controllable, some state of system (A,B) cannot be driven to zero in $\nu-1$ periods. If one considers the pair (A,B), then a further result (Kabamba, 1985) is that the number of periods required to perform the controllability transition is at most the controllability index (Chen, 1970) of $(\Phi(t+T,t),\ W_c(t,t+T))$.

2. In this subsection, only the controllability notion has been taken into consideration. Needless to say that, since $X_r(t) = X_c(t)$, analogous results can be given for reachability.

3. A periodic output equation can be added to the state equation (1):

$$y(t) = C(t)\ x(t)$$

with $C : R \to R^{pxn}$ continuous, real and T-periodic:

$$C(t+T) = C(t) \quad , \quad \forall t.$$

Then, the notions of state observability and reconstructibility can be introduced. As is well known, Kalman (1969), Chen (1970), these notions deal with the possibility of distinguishing a free motion starting from or ending in a given state from the free motion starting from or ending in the origin of the state-space. Obviously, this last free motion results in the zero output function. The observability and reconstructibility of (A,C) are not explicitely considered here. The reason is that the observability and reconstructibility properties of periodic systems can be derived from the ones concerning reachability and controllability via the duality theory, Kalman (1969), Chen (1970). Indeed, the observability [reconstructibility] of (A,C) is equivalent to the reachability [controllability] of the *dual pair* (A', C').

3. Structural Properties of Discrete-time Periodic Systems

3.1 Discrete-time linear periodic systems

Turning now to discrete-time systems, consider the difference equation

$$x(t+1) = A(t)x(t) + B(t)\,u(t) \qquad (5.a)$$

where $A : Z \to R^{nxn}$ and $B : Z \to R^{nxm}$ and (for an integer T)

T-periodic:

$$A(t+T) = A(t) \quad ; \quad B(t+T) = B(t) \quad , \forall t \qquad (5.b)$$

Since A may be singular at certain time points, system (5) may not be reversible. This is a major difference between discrete -time and continuous-time systems. As a consequence of non-reversibility, the trajectory portrait in discrete-time can be quite different than in continuous-time. Precisely, in continuous-time there is one and only one free motion passing through any given event (x,t). In discrete-time, there may exist events where two or more free motions end and events where no free motion end. If (x,t) is an event with no free motion ending in it, x is named *new-born* state at time t, Bittanti and Bolzern (1986).

The system transition matrix is now given by

$$\Phi(t,\tau) = A(t-1)\, A(t-2) \ldots A(\tau).$$

The spectrum of the matrix $\Phi(t+T,t)$ is independent of t. Indeed, for any pair $(\tau,\ t \in [0,T-1])$, the monodromy matrices can be expressed in the form $\Phi(t+T,t) = FG$ and $\Phi(\tau+T,\tau) = GF$. Hence, if λ is a nonzero eigenvalue of $\Phi(t+T,t)$, i.e. $FGx = \lambda x$, $x \neq 0$, then $GFy = \lambda y$, where $y = Gx$. Notice that, since $\lambda \neq 0$ and $x \neq 0$, x cannot belong to the null-space of G. Hence $y \neq 0$, so that λ is an eigenvalue of GF as well. Therefore, all the nonzero eigenvalues of $\Phi(t+T,t)=FG$ are eigenvalues of $\Phi(\tau+T,\tau) = GF$. By reversing the role of $\Phi(t+T,t)$ and $\Phi(\tau+T,\tau)$ in the above argument, it also follows that all the nonzero eigenvalues of $\Phi(\tau+T,\tau)$ are eigenvalues of $\Phi(t+T,t)$. This implies that the nonzero eigenvalues of $\Phi(t+T,t)$ and $\Phi(\tau+T,\tau)$ do coincide, so demonstrating the time invariance of the spectrum Σ.

Some eigenvalues of $\Phi(t+T,t)$ may be zero. The symbol $\sum_0$ will denote the set of the nonzero elements of $\sum$.

Although the characteristic polynomial of $\Phi(t+T,t)$ does not change with t, the minimal polynomial may depend on t. The degree of the minimal polynomial will be denoted by ν_t. Finally, note that the determinant of $\Phi(t+T,t)$ may not be positive.

3.2 Structural properties

The definitions of reachability and controllability given above apply to discrete-time systems as well. When considering the structural properties, the only care to be taken in discrete-time concerns the definition of reconstructibility. This is due to the peculiar phase portrait of nonreversible systems. Since the attention is mainly focused here on reachability and controllability, this aspect will not be discussed in this paper. The interested reader is referred to Bittanti and Bolzern (1986).

3.3 Grammian matrices

The only Grammian matrix which can be defined in general is the reachability one:

$$W_r(\tau,t) = \sum_{\tau+1}^{t} {}_j \, \Phi(t,j)B(j-1)B(j-1)'\Phi(t,j)'. \tag{6}$$

For reversible systems, the backward transition matrix $\Phi(\tau,t) = \Phi(t,\tau)^{-1}$, $t \geq \tau$, can be defined. Then, the controllability Gramian matrix is given by:

$$W_c(t,\tau) = \sum_{t+1}^{\tau} {}_j \, \Phi(t,j)B(j-1)B(j-1)'\Phi(t,j)'.$$

As in continuous-time, $X_r(\tau,t) = R\left[W_r(\tau,t)\right]$. Moreover, for reversible systems, $X_c(t,\tau) = R\left[W_c(t,\tau)\right]$.

In particular, system (5.a) is reachable at time t if and only if, for some $\tau<t$, the matrix $W_r(\tau,t)$ is full rank. For time-invariant systems, the system reachability can also be tested by resorting to the PBH condition. This coincides with the PBH condition for continuous-time systems (Sect. 2.4).

3.4 Five structural properties of discrete-time periodic systems

Property A is false for periodic discrete-time systems. The only conclusion which holds in general is that (Bittanti and Bolzern, 1984,a, Lemma 2)

$$X_r(t) \subseteq X_c(t) \quad , \quad \forall t.$$

Furthermore, only the dimension of the controllability subspace is time-invariant (Bittanti and Bolzern, 1984,a, Lemma 2)

$$\dim \; X_c(t) = \text{const.} \tag{7}$$

The following simple example shows that the dimension of the reachability subspace may change in time.

Example 2

Consider the scalar system

$$x(t+1) = a(t)\; x(t) + b(t)\; u(t)$$

$$a(t) = b(t) = \begin{cases} 0 \; , & t \text{ even} \\ 1 \; , & t \text{ odd.} \end{cases}$$

Then:

$$X_r(t) = \begin{cases} R^1 & , \ t \text{ even} \\ \{0\} & , \ t \text{ odd.} \end{cases} \square$$

As for Property C, from (7) it follows that, if system (5) is controllable at a certain time point, it is controllable at any time point. It is apparent from Example 2 that the same conclusion is false for reachability.

In order to guarantee the system reachability at any time point, it is necessary to require the reachability at each time point of a suitable set Δ, which reduces to any singleton in $[0, T-1]$ for reversible systems only. Precisely, let

$$\Delta = \begin{cases} \text{any } t \in [0, T-1] \text{ if the system is reversible} \\ \{t \in [0,T-1] \text{ such that } \det A(t-1)=0\} & , \text{ otherwise.} \end{cases}$$

Then, by analyzing the reachability Gramian, in Bittanti and Bolzern (1985,b) it is proven that system (5) is reachable if it is reachable for each $t \in \Delta$.

As far as the reachability and controllability interval length, in Bittanti and Bolzern (1984,a, Lemma 1) the following result is derived by considerations of geometric kind:

$$X_r(t) = X_r(t-nT,t) \quad , \forall t$$

$$X_c(t) = X_c(t, t+nT) \quad , \forall t.$$

Obviously, this entails the discrete-time version of Brunovsky Theorem. Another result concerning the reachability and controllability interval length, given in Bittanti and Bolzern (1985,b), can be stated as follows:

$$X_r(t) = R^n \implies X_r(t) = X_r(t-\nu_t T, t) \qquad (8.a)$$

$$X_c(t) = R^n \implies X_c(t) = X_c(t, t+\nu_t T), \qquad (8.b)$$

where it is recalled that ν_t is the degree of the minimal polynomial of $\Phi(t+T,t)$.

Remarks

4. For nonreversible systems, ν_t may depend upon t. This fact may give raise to an interesting paradox.
   Consider a controllable system with, say, $\nu_1=3$ and $\nu_2=1$. Then, starting from t=1, any state can be driven to zero in at most 3 periods. However, since any state can also be driven to zero in an interval no longer than T starting from t=2, the stronger conclusion would follow that, starting from t=1, any state can be driven to zero in at most 2 periods.
   However, this paradox resolves readily, since it can be shown (Bittanti and Bolzern, 1985,b) that

$$|\nu_t - \nu_\tau| \leq 1 \quad , \forall \tau, t$$

5. The discrete-time version of the result mentioned in Remark 1 can be found in Bittanti, Colaneri, De Nicolao (1986).

Precisely, let μ_{rt} be the reachability index of the pair $(\Phi(t+T,t), W_r(t,t+T))$ and denote by $\mu_{ct} = \max(\mu_{rt}, \mu_z)$ where μ_z is the dimension of the largest Jordan block of the controllable and unreachable part of system $(\Phi(t+T,t), W_r(t,t+T))$. Then

$$X_r(t) = X_r(t-\mu_{rt}T, t)$$

$$X_c(t) = X_c(t, t+\mu_{ct}T)$$

From this, it follows that conclusions (8) hold true even if the assumptions $X_r(t)=R^n$ and $X_c(t)=R^n$ are removed.

6. The impossibility of defining in general a controllability Grammian leads to the question of working out a controllability test of periodic systems based on the reachability Gramian. A test of this type can be found in Bittanti and Bolzern (1985,b). □

The discrete-time version of the spectral characterizations of the structural properties presented in Sect. 2.5 can be stated as follows (Bittanti and Bolzern, 1985,b, Sect. 6).

Theorem 3

System (5) is reachable [controllable] at time t if and only if the following H-condition at t is satisfied.

H-condition at t (discrete-time)

(i) For reachability:

Given the time point t, the matrix

$$\left[sI - \Phi(t+T,t) \qquad W_r(t,t+T)\right] \tag{9}$$

is full rank on Σ.

(ii) For controllability :

Given the time point t, the matrix (9) is full rank on Σ_0.

Remark

7. As in continuous-time, if the H-controllability condition is satisfied at a time point, it is satisfied at each time point. As it follows from the previous discussion on the reachability properties of periodic systems, the same conclusion does not apply to the H-reachability condition.

4. Kalman Canonical Decomposition

A fundamental result in System Theory is that any time-invariant and continuous-time system can be decomposed in 4 parts, i.e. the controllable-observable, controllable-unobservable, uncontrollable-observable and uncontrollable-unobservable parts.

For continuous-time systems, this result is extended to periodic systems in the appendix of (Bittanti and Bolzern, 1985,c). The proof is based on the time-invariance of the rank of the Gramians, which corresponds to the time invariance of the dimensions of the structural subspaces. The result says that, by means of a T-periodic state transformation, any T-periodic

system can be decomposed into the 4 parts of the Kalman canonical decomposition.
As discussed in the previous section, the reachability and controllability subspaces may not coincide for a discrete-time periodic system. Dually, the observability and reconstructibility subspaces may not coincide too. Therefore, four canonical decompositions sbould have to be considered, i.e. the canonical decompositions based on the pairs (reachability, observability), (controllability, observability), (reachability, reconstructibility) and (controllability, reconstructibility). To see which one of these decompositions can actually be used for discrete-time periodic systems, note that the reachability subspace and the dual observability subspace may have time-varying dimensions (Sect. 3.4). Consequently, only the (controllability, reconstructibility) decomposition is a candidate for a canonical decomposition of general validity. This is why, contrary to the continuous-time case, the theory cannot be based on the Gramian matrices. Indeed, as already observed, only the reachability and observability Gramians can be defined for discrete-time systems.

In (Bittanti and Bolzern, 1984,b and 1986), a theory for the Kalman canonical decomposition of *any* time-varying and discrete -time system is worked out. Letting $X_a(t)$ be either the reachability or the controllability subspace at t and $\bar{X}^{\alpha}(t)$ be either the unobservability or unreconstructibility subspace, it is proven that an (a,α) canonical decomposition exists if the following three conditions are met with.

(i) dim $X_a(t)$ = const.

(ii) dim $\bar{X}^{\alpha}(t)$ = const.

(iii) dim $X_a(t) \cap \bar{X}^{\alpha}(t)$ = const.

(i)-(iii) are named *dimension-invariance condition.*

In (Bittanti and Bolzern, 1986), it is also shown that, for periodic systems, the dimension-invariance condition is verified by taking a= controllability and α=reconstructibility. In conclusion, discrete-time periodic systems can be canonically decomposed by making reference to the pair (controllability, reconstructibility). By focusing on periodic systems only, this result is also derived in Grasselli (1984).

5. Extended Structural Properties

The notions of stabilizability and detectability are here called extended structural properties. Since the detectability results can be derived from the ones relative to stabilizability by duality, only stabilizability is considered in this section.

As a concise introduction to the stabilizability concept, consider the time invariant system:

$$\dot{x}(t) = Ax(t) + B\,u(t) \tag{10}$$

or

$$x(t+1) = Ax(t) + B\,u(t)\ . \tag{11}$$

A classical result of System Theory, see e.g. Kalman (1969) is that, if (A,B) is reachable, then the system can be stabilized by a suitable feedback control law. Stated differently, there exists a matrix $K \in R^{mxn}$ such that

$$\dot{x}(t) = Ax(t) + B\,u(t)$$
$$u(t) = Kx(t)$$

or

$$x(t+1) = Ax(t) + B\,u(t)$$
$$u(t) = Kx(t)$$

is asymptotically stable respectively. Thus, the system reachability is a sufficient condition for the existence of a stabilizing feedback control law. However, it is not necessary. This leads to the notion of stabilizability. Precisely, according to a classical definition by Wonham (1968), a system is stabilizable whenever there exists a stabilizing feedback control law.

While the PBH condition for reachability is the same in continuous or discrete-time, the PBH condition for the stabilizability of time-invariant systems is different in continuous or discrete-time. Precisely, system (10) [(11)] is stabilizable if and only if the matrix

$$\begin{bmatrix} sI - A & B \end{bmatrix}$$

is full rank for all eigenvalues of A with nonnegative real part [with modulus greater than or equal to 1].

The main characterizations of the stabilizability notion for continuous-time periodic system can be summarized as follows

(Bittanti and Bolzern, 1985,b) and also (Bittanti and Bolzern, 1984,c) and (Shayman, 1984). Recall that $\sum_1$ is the set of the eigenvalues of $\Phi(t+T,t)$ not belonging to the open unit disk (unstable part of the spectrum).

Theorem 4

The following statements are equivalent to each other.

(a) There exists a T-periodic matrix $K: R \rightarrow R^{mxn}$ such that

$$\dot{x}(t) = [A(t) - B(t)\ K(t)]\ x(t)$$

is asymptotically stable.

(b) The uncontrollable part of system (1) is asymptotically stable.

(c) For at least a time point $t \in [0,T]$, the matrix

$$[sI - \Phi(t+T,t) \qquad W_c(t,t+T)]$$

is full rank on $\sum_1$. □

Characterizations (a) and (b) are direct extensions to periodic systems of analogous characterizations for time-invariant systems. (c) is a natural extension of the H-controllability condition (Sect. 2.5). In view of the discrete-time version of the PBH test, (c) is equivalent to saying that system (1) is stabilizable if and only if the discrete-time pair defined by $(\Phi(t+T,t),\ W_c(t,t+T))$ is stabilizable.

The discrete-time version of this theorem is given in (Bolzern,

1986) and can be stated as follows.

Theorem 5

The following statements are equivalent:

(a) There exists a T-periodic matrix $K : Z \to R^{mxn}$ such that

$$x(t+1) = \left[A(t) - B(t)\; K(t)\right] x(t)$$

is asymptotically stable.

(b) The uncontrollable part of system (1) is asymptotically stable.

(c) For at least one time point $t \in [0,T]$, the matrix

$$\left[sI - \Phi(t+T,t) \qquad W_r(t,t+T)\right]$$

is full rank on $\sum_1$. □

The modal characterizations of stabilizability (point (c) of Thm. 4 and 5) are extensions of the H-conditions previously introduced. Needless to say, if the matrix $[sI - \Phi(t+T,t)$ $W_c(t,t+T)]$ is full rank on $\sum_1$ at a given t, then the same matrix is full rank on $\sum_1$ at any time. An analogous statement holds for discrete-time systems.

6. Stochastic Linear Periodic Systems

This section is devoted to the study of linear periodic systems subjet to inputs which are stochastic processes of periodic type. Precisely, focusing in this Section on the continuous-time case only, the system taken into consideration is

$$dx(t) = A(t)\, x(t)\, dt + B(t)\, dv(t) \tag{12}$$

where v is an m vector valued stochastic process characterized as follows:

Let

$$m(t) := E\left[v(t)\right] ;$$

Then

$$v(t) = m(t) + z(t) \tag{13}$$

where z satisfies the stochastic differential equation

$$dz(t) = q(t)\, dw(t) \quad , \quad z(0) = 0. \tag{14}$$

In (14), w is an m-dimensional standard Wiener process, while q is T-periodic and continuous. Moreover, it is also assumed that function m in (13) is T-periodic, continuous and piecewise continuously differentiable.

Therefore, (12) is a stochastic differential equation of the form

$$dx(t) = \left[A(t)\, x(t) + B(t)\, \dot{m}(t)\right] dt + \eta(t)\, dw(t) \tag{15}$$

with

$$\eta(t) := B(t) q(t)$$

For a given random vector x(0) independent of $\{w(t),\ t \geq 0\}$, the meaning of (15) is precisely that

$$x(t) = x(0)+ \int_0^t \left[A(\sigma)x(\sigma)+B(\sigma)\dot{m}(\sigma)\right]d\sigma+ \int_0^t \eta(\sigma)dw(\sigma), \qquad (16)$$

the stochastic integral being understood in the sense of Ito (Wong and Hajek, 1985, Ch. 4). The solution of (15), or its equivalent integral version (16), is given by

$$x(t)= \Phi(t,0)x(0)+\int_0^t \Phi(t,\sigma)B(\sigma)\dot{m}(\sigma)d\sigma+\int_0^t \Phi(t,\sigma)\eta(\sigma)dw(\sigma) \qquad (17)$$

as follows readily from the Ito differential rule.

Should x(0) be Gaussian, say with expected value z_0 and covariance matrix Γ_0, then (17) defines a Gaussian process, with expected value

$$z(t) = \Phi(t,0)\ z_0+ \int_0^t \Phi(t,\sigma)B(\sigma)\dot{m}(\sigma)d\sigma \qquad (18)$$

and covariance matrix

$$\Gamma(t) = \Phi(t,0)\ \Gamma_0\ \Phi(t,0)' +\int_0^t \Phi(t,\sigma)Q(\sigma)\Phi(t,\sigma)'d\sigma \qquad (19)$$

where

$$Q(t) = \eta(t)\ \ \eta(t)'$$

represents the covariance matrix of the noise entering (15). From (18) and (19), it is well known (Brockett, 1970) that

$$\dot{z}(t) = A(t)\; z(t) + B(t)\; m(t) \tag{20}$$

$$\dot{\Gamma}(t) = A(t)\Gamma(t) + \Gamma(t)A(t)' + B(t)Q(t)B(t)' \tag{21}$$

with the initial conditions

$$z(0) = z_0 \quad , \qquad \Gamma(0) = \Gamma_0 .$$

The process covariance function,

$$\gamma(t,\tau) := E\left[(x(t) - z(t))\;(x(\tau) - z(\tau))'\right],$$

possesses the following properties

$$\gamma(t,\tau) = \gamma(\tau,t)'$$

and, for $t > \tau$,

$$\gamma(t,\tau) = \Phi(t,\tau)\;\Gamma(\tau). \tag{22}$$

Eq.(22) can be derived from (17) by simple computations.

Since A, B, m and Q are T-periodic, it is natural to investigate the existence of periodic solutions of (20) and (21).
If such is the case,

$$\gamma(t+T,\,\tau+T) = \Phi(t+T,\,\tau+T)\;\Gamma(\tau+T) = \Phi(t,\tau)\;\Gamma(\tau) = \gamma(t,\tau)$$

in view of (2).

A stochastic process with T-periodic expected value and covariance function satisfying

$$\gamma(t+T,\ \tau+T) = \gamma(t,\tau) \quad , \quad \forall t,\tau$$

is called *cyclo-stationary* (Gardner and Franks, 1975). Processes of this type find numerous applications in various areas. They can be used for the modelling of seasonal time series or to describe uncertain signals of periodic type. For an illustrative example, see (Bittanti and Hernandez, 1986).

In conclusion, the problem is to find initial conditions (22) for which there are periodic solutions to (20) and (21), that is, solutions satisfying

$$z(T) = z(0) \tag{23}$$

$$\Gamma(T) = \Gamma(0) \tag{24}$$

respectively. More precisely, for the Lyapunov equation (21), the problem is to find a T-periodic symmetric solution which is positive semidefinite at each time point.

As for the expected value, it is easy to see that, if no characteristic multiplier is equal to 1, then (20) admits a unique solution satisfying (23).

Consider now the Lyapunov equation (21) and, given a nxn real and symmetric matrix $\bar{\Gamma}$, let $\Gamma_\tau(t)$ be a solution such that $\Gamma_\tau(\tau) = \bar{\Gamma}$. It is well known that, for any $\bar{\Gamma}$, (21) has a unique solution $\Gamma_\tau(t)$, $-\infty < t < +\infty$, see e.g. Brockett (1970). Let $\tilde{B}$ be a T-periodic matrix such that

$$\tilde{B}(t)\ \tilde{B}(t)' = B(t)\ Q(t)\ B(t)'$$

and denote by $\tilde{W}_r(\tau,t)$ the reachability Grammian matrix of $(A,\tilde{B})$, i.e. (see (3)),

$$\tilde{W}_r(\tau,t) = \int_{\tau}^{t} \Phi(t,\sigma)\tilde{B}(\sigma)\tilde{B}(\sigma)'\Phi(t,\sigma)'d\sigma \tag{25}$$

For $t>\tau$, the solution of the Lyapunov equation is given by

$$\Gamma_{\tau}(t) = \Phi(t,\tau)\ \bar{\Gamma}\ \Phi(t,\tau)' + \tilde{W}_r(\tau,t). \tag{26}$$

Setting now $\tau=0$, $t=T$ and imposing the periodicity constraint (24), the following equation is obtained:

$$\bar{\Gamma} = \Phi(T,0)\ \bar{\Gamma}\ (T,0)' + \bar{W}.$$

This is the discrete-time algebraic Lyapunov equation. It can be shown (Graham, 1981) that, if the characteristic multipliers lie within the unit circle, this equation admits a unique solution.

From these results, it follows that: if the system is asymptotically stable, then both (20) and (21) admit a unique T-periodic solution.

As a matter of fact, under the assumption of asymptotic stability, the following can be shown to hold true (Bittanti, Bolzern and Colaneri, 1984):
Consider the solution $\Gamma_{\tau}(t)$ of (21) such that $\Gamma_{\tau}(\tau) = \bar{\Gamma}$. Then $\Gamma_{\tau}(t)$ converges to the periodic solution of (21) as $\tau\to-\infty$, for whichever $\bar{\Gamma}$. In particular, taking $\bar{\Gamma}=0$, (26) entails that the $\tilde{W}_r(-\infty,t)$ *is* the T-periodic solution. Moreover,

in view of (25), it is also apparent that this solution is positive semidefinite (at each t). In fact, should $(A,\tilde{B})$ be reachable, the solution is obviously positive definite (at each t).

This last conclusion is part of the so-called Periodic Lyapunov Lemma, which can be stated as follows (Bittanti, Bolzern and Colaneri, 1985).

Theorem 6

The system is asymptotically stable if and only if, for any $\tilde{B}$ such that $(A,\tilde{B})$ is reachable, there exists a T-periodic positive definite solution of the Lyapunov equation (21). □

An extended version of this lemma can be given under the assumption that $(A(t), \tilde{B}(t))$ be stabilizable only.

Theorem 7

The system is asymptotically stable if and only if, for any $\tilde{B}$ such that $(A,\tilde{B})$ is stabilizable, there exists a T-periodic positive semidefinite solution of the Lyapunov equation (21). □

Theorem 7 is proven in (Bittanti, Bolzern and Colaneri, 1985) by means of a decomposition technique. Precisely, the Lyapunov equation is decomposed into three subequations corresponding to the reachability canonical decomposition of $(A,\tilde{B})$.

One could wonder whether the Lyapunov equation may admit a T-periodic positive semidefinite solution even if the system is not stable. In case the system is not asymptotically stable, matrix $\Phi(T,0)$ has some eigenvalues on or outside the unit circle.

If a characteristic multiplier lies on the unit circle, and the pair $(A,\tilde{B})$ is stabilizable, then (21) does not admit any T-periodic solution, (see (Wimmer and Ziebur, 1975) and (Bittanti and Colaneri, 1986, Thm. 2(a)). Suppose now that, say, p characteristic multipliers have modulus greater than 1, while the remaining n-p ones have modulus lower than 1. Then, in (Shayman, 1984) and (Bittanti, Colaneri, 1986), it is shown that, if $(A,\tilde{B})$ is reachable or stabilizable, the T-periodic solution of (21) (if any) has p negative eigenvalues for each time-points. The remainign n-p ones are all positive if $(A,\tilde{B})$ is reachable or nonnegative if $(A,\tilde{B})$ is stabilizable.

Since it is obvious that only the positive semidefinite solutions of (21) correspond to a cyclostationary solution of (12), the conclusion is the following: Assume that $(A,\tilde{B})$ is stabilizable. Then, if the system is not asymptotically stable, eq.(12) admits no cyclostationary solution.

The analysis of the discrete-time periodic Lyapunov equation is currently underway and partially reported in (Bolzern and Colaneri, 1986).

Acknowledgment

The author is grateful to Professors Guido Guardabassi and Diego Bricio Hernandez for their helpful and stimulating comments.

References

Bailey, J.E. (1973): Periodic Operation of Chemical Reactors: A Review. Chem. Eng. Commun. 1, 111-124.

Bekir, E. and R.S. Bucy (1976): Periodic Equilibria for Matrix Riccati Equations. Stochastics 2, 1-104.

Belevitch, V. (1968): Classical Network Theory. Holden Day, San Francisco.

Bernstein, D.S and E.G. Gilbert (1980): Optimal Periodic Control: The Π Test Revisited. IEEE Trans. Automatic Control AC-25, 673-684.

Bittanti, S. and P. Bolzern (1984,a): Can the Kalman Canonical Decomposition be performed for a Discrete-time Linear Periodic System? 1st Latin American Conference on Automatica, Campina Grande, Brazil, 449-453.

Bittanti, S. and P. Bolzern (1984,b): Canonical Decomposition and Discrete-time Linear Systems. 23rd Conference of Decision and Control, Las Vegas, U.S.A., 1737, 1738.

Bittanti, S. and P. Bolzern (1984,c): Four Equivalent Notions of Stabilizability of Periodic Linear Systems. 3rd American Control Conference, San Diego, U.S.A., 1321-1323.

Bittanti, S. and P. Bolzern (1985,a): Reachability and Controllability of Discrete-time Linear Systems. IEEE Trans. Automatic Control 30, 399-491.

Bittanti, S. and P. Bolzern (1985,b): Discrete-time Linear Periodic Systems: Grammian and Modal Criteria for Reachability and Controllability. International J. Control 41, 899-928.

Bittanti, S. and P. Bolzern (1985,c): Stabilizability and Detectability of Linear Periodic Systems. Systems and Control Letters 6, 141-145. Plus Addendum, to appear in Systems and Control Letters (1986), 7, 73.

Bittanti, S. and P. Bolzern (1986): On the Structure Theory of Discrete-time Linear Systems. International J. Systems Science, 17, 33-47.

Bittanti, S., P. Bolzern and P. Colaneri (1984): Stability Analysis of Linear Periodic Systems via the Lyapunov Equation. 9th IFAC World Congress, Budapest, 8, 169-172.

Bittanti, S., P. Bolzern and P. Colaneri (1985): The Extended Periodic Lyapunov Lemma. Automatica 5, 603-605.

Bittanti, S., P. Bolzern, P. Colaneri and G. Guardabassi (1983): H and K-Controllability of Linear Periodic Systems. 22nd Conference on Decision and Control, S. Antonio, U.S.A., 1376-1379.

Bittanti, S. and P. Colaneri (1986): Lyapunov and Riccati Equations: Periodic Inertia Theorems. IEEE Trans. Automatic Control (to appear).

Bittanti, S., P. Colaneri and G. De Nicolao (1986): Discrete-time Periodic Systems: a note on the Reachability and Controllability interval length. Centro Teoria Sistemi, Politecnico di Milano, Int. Rep. 86-003.

Bittanti, S., P. Colaneri and G. Guardabassi (1984): H-Controllability and Observability of Linear Periodic Systems. SIAM J. Control and Optimization 22, 889-893.

Bittanti, S., G. Fronza and G. Guardabassi (1973): Periodic Control: A Frequency Domain Approach. IEEE Trans. Automatic Control 18, 33-38.

Bittanti, S., G. Guardabassi, C. Maffezzoni and L. Silverman (1978): Periodic Systems: Controllability and the Matrix Riccati Equation. SIAM J. Control and Optimization 16, 37-40.

Bittanti, S. and D.B. Hernandez (1986): The Simple Pendulum as an Illustrative Example of the Periodic Control Problem. Centro Teoria dei Sistemi, Politecnico di Milano, Int. Rep. 86-010.

Bolzern, P. (1986): Criteria for Reachability, Controllability and Stabilizability of Discrete-time Linear Periodic Systems. V Polish-English Seminar on Real-Time Process Control, Warsaw, Poland.

Bolzern, P. and P. Colaneri (1986): Existence and Uniqueness Conditions for the Periodic Solutions of the Discrete-time Periodic Lyapunov Equation. Centro Teoria dei Sistemi, Politecnico di Milano, Int. Rep. 86-011.

Brockett, R.W. (1970): Finite Dimensional Linear Systems. J. Wiley and Sons.

Brunovsky, P. (1969): Controllability and Linear Closed loop Controls in Linear Periodic Systems. J. Differential Equations 6, 296-313.

Chen, C.T. (1970): Introduction to Linear System Theory. Holt, Rinehart and Winston.

Colonius, F. (1985,a): Optimality for Periodic Control of Functional Differential Systems. J. Mathematical Analysis and Applications (to appear).

Colonius, F. (1985,b): The High Frequency Pi-Criterion for Retarded Systems. IEEE Trans. Automatic Control 11, 1045-1048.

DaPrato, G. (1984): Periodic Solutions of Infinite Dimensional Riccati Equations. Rendiconti Accademia Nazionale dei Lincei, (to appear).

Dorato, P. and A.H. Levis (1971): Optimal Linear Regulators: the Discrete-time Case. IEEE Trans. Automatic Control 6, 613-620.

Dorato, P. and H.K. Knudsen (1979): Periodic Optimization with Applications to Solar Energy Control. Automatica 15, 673-679

Gardner, W.A. and D.E. Franks (1975): Characterization of Cyclo-stationary Random Processes. IEEE Trans. Information Theory 21, 1-24.

Gilbert, E.G. (1977): Optimal Periodic Control: A General Theory of Necessary Conditions. SIAM J. Control and Optimization 15, 717-746.

Gilbert, E.G. and D.T. Lyons (1981): The Improvement of Aircraft Specific Range by Periodic Control. AIAA Guidance and Control Conference, Albuquerque.

Graham, A. (1981): Kronecker Products and Matric Calculus with Applications. Ellis Horwood Limited, Chichester.

Grasselli, O.M. (1984): A Canonical Decomposition of Linear Periodic Discrete-time Systems. International J. Control 40, 201-214.

Guardabassi, G. (1971): Optimal Steady State Versus Periodic Control. Ricerche di Automatica 2, 240-252.

Guardabassi, G. (1976): The Optimal Periodic Control Problem. Journal A 17, 75-83.

Halanay, A.(1966): Differential Equations. Academic Press, New York.

Hautus, M.L.J. (1969): Controllability and Observability Conditions of Linear Autonomous Systems. Indag. Math. 72 443-448.

Hernandez, V. and L. Jodar (1985): Boundary Problems and Periodic Riccati Equations. IEEE Trans. Automatic Control 11, 1131-1133.

Hewer, G.A. (1975): Periodicity, Detectability and the Matrix Riccati Equation. SIAM J. Control 13, 1235-1251.

Horn, F.J.M. and R.C. Lin (1967): Periodic Processes: A Variational Approach. Ind. Eng. Chem. Process Des. Dev. 6, I, 21-30.

Horn, F.J.M. and J.E. Bailey (1968): An Application of the Theorem of Relaxed Control to the Problem of Increasing Catalyst Selectivity. J. Optimization Theory and Applications 2, 441-449.

Houlihan, S.C., E.M. Cliff and H.J. Kelley (1982): Study of Chattering Cruise, Journal Aircraft 19, 119-124.

Johnson, C.D. (1966): Invariant Hyperplanes for Linear Dynamical Systems. IEEE Trans. Automatic Control 11, 113-116.

Kabamba, P.T. (1985): Monodromy Eigenvalue Assignment in Linear Periodic Systems. 24th Conference on Decision and Control, Ft. Lauderdale, U.S.A., 177, 178.

Kalman, R.E. (1969): Theory of Regulators for Linear Plants. In: Kalman R.E., P.L. Falb and M.A. Arbib: Topics in Mathematical System Theory. McGraw-Hill Co., New York.

Kano, H. and T. Nishimura (1979): Periodic Solutions of Matrix Riccati Equations with Detectability and Stabilizability. International J. Control 29, 471-487.

Kern, G. (1980): Linear Closed-loop Control in Linear Periodic Systems with Application to Spin-stabilized Bodies. International J. Control 31, 905-916.

Khandelwal, D.N., J. Sharma and L.M. Ray (1979): Optimal Periodic Maintenance of a Machine. IEEE Trans. Automatic Control 24, 513.

Khargonekar, P.P., K. Poolla and A. Tannenbaum (1985): Robust Control of Linear Time-invariant Plants Using Periodic Compensation. IEEE Trans. Automatic Control 11, 1088-1098.

Kono, M. (1980): Eigenvalue Assignment in Linear Periodic Discrete-time Systems. International J. Control 1, 149-158.

Maffezzoni, C. (1974): Hamilton-Jacobi Theory for Periodic Control Problems. J. Optimization Theory and Applications 14, 21-29.

Markus, L. (1973): Optimal Control of Limit Cycles or what Control Theory can do to Cure a Heart Attack or to Cause One. Symposium on Ordinary Differential Equations, Minneapolis, Minnesota (1972). W.A. Harris, Y. Sibuya, eds., Springer-Verlag, Berlin.

Matsubara, M., N. Nishimura, N. Watanabe and K. Onogi (1981): Periodic Control Theory and Applications. Research Reports of Automatic Control Laboratory Vol. 28, Faculty of Engineering, Nagoya University.

Matsubara, M. and K. Onogi (1978): Stabilized Suboptimal Periodic Control of a Chemical Reactor. IEEE Trans. Automatic Control 23, 1005-1008.

Meyer, R.A. and C.S. Burrus (1976): Design and Implementation of Multirate Digital Filters. IEEE Trans. Acoustics, Speech and Signal Processing 1, 53-58.

Nistri, P. (1983): Periodic Control Problems for a Class of Nonlinear Periodic Differential Systems. Nonlinear Analysis, Theory, Methods and Applications 7, 79-90.

Noldus, E. (1975): A Survey of Optimal Periodic Control of Continuous Systems. Journal A 16, 11-16.

Onogi, K. and M. Matsubara (1980): Structure Analysis of Periodically Controlled Chemical Processes. Chem. Eng. Sci. 34, 1009-1019.

Popov, V.M. (1973): Hyperstability of Control Systems. Springer, Berlin.

Schädlich, K., U. Hoffmann and H. Hofmann (1983): Periodical Operation of Chemical Processes and Evaluation of Conversion Improvements. Chemical Engineering Science 38, 1375-1384.

Shayman, M.A. (1984): Inertia Theorems for the Periodic Lyapunov Equation and Periodic Riccati Equation. Systems and Control Letters 4, 27-32.

Shayman, M.A. (1985): On the Phase Portrait of the Matrix Riccati Equation Arising from the Periodic Control Problem. SIAM. J. Control and Optimization 23, 717-751.

Sincic, D. and J.E. Bailey (1978): Optimal Periodic Control of Variable Time-delay Systems. International J. Control 27, 547-555.

Speyer, J.L. (1973): On the Fuel Optimality of Cruise, J. Aircraft 10, 763-764.

Speyer, J.L. (1976): Non-optimality of Steady-state Cruise for Aircraft. AIAA Journal 14, 1604-1610.

Speyer, J.L. and R.T. Evans (1984): A Second Variational Theory of Optimal Periodic Processes. IEEE Trans. Automatic Control 29, 138-148.

Valko', P. and G.A. Almasy (1982): Periodic Optimization of Hammerstein-type Systems. Automatica 18, 245-148.

Watanabe, N., Y. Nishimura and M. Matsubara (1976): Singular Control Test for Optimal Periodic Control Problems. IEEE Trans. Automatic Control 21, 609-610.

Watanabe, N., K. Onogi and M. Matsubara (1981): Periodic Control of Continuous Stirred Tank Reactors - I, Chem. Eng. Sci. 36, 809-818, II ibid. 37, 745-752.

Watanabe, N., H. Kurimoto and M. Matsubara (1984): Periodic Control of Continuous Stirred Tank Reactors - III,Case of multistage reactors. Chem. Eng. Sci. 39, 31-36.

Wimmer, H.K. and A.D. Ziebur (1975): Remarks on Inertia Theorems for Matrices. Czechoslovak Mathematical Journal 25, 556-561.

Wong, E. and B. Hajek (1985): Stochastic Processes in Engineering. Springer-Verlag, Berlin.

Wonham, W.M. (1968): On a Matrix Riccati Equation for Stochastic Control. SIAM Journal Control 6, 681-698.

Yakubovich, V.A. and V.M. Starzhinskii (1975): Linear Differential Equations with Periodic Coefficients. J. Wiley, New York.

Chapter 6

Numerical Problems in Linear System Theory

Daniel Boley and Sergio Bittanti

1. Introduction

The analysis of multivariable control systems involves the computation of matrix problems, ranging from linear systems to matrix rank and eigenvalues. In general, the techniques to be used on a computer are not the same as those used for hand computations. In this work, we outline a few techniques for computer calculations, illustrate why they are useful, and give examples of where numerical problems arise in system theory.

We begin with a review of the simpler decompositions used to solve systems of linear equations and related problems, then go on to more sophisticated methods for eigenvalue and rank computation (Schur and Singular Value Decompositions). We conclude with a few applications of these concepts to the analvsis of time-invariant linear systems.

2. Review of Simpler Computational Methods

2.1 - LU decomposition

We begin by introducing the concept of a matrix decomposition; that is, we try to reduce a matrix A to the product of several simpler matrices, from which we can calculate whatever it is we would like to calculate.

The first example is the LU decomposition, in which we decompose a matrix A into the product A = LU, where L, U are lower, upper triangular, respectively. This decomposition is computed using Gaussian Elimination. To see this, it is best to use an example.

Consider

$$A = \begin{bmatrix} 3 & 1 & 6 \\ 1 & 1 & 1 \\ 2 & 1 & 5 \end{bmatrix} . \tag{1}$$

In Gaussian Elimination, the first step is to add multiples of row 1 to rows 2 and 3. This can be accomplished by multiplying A on the left by the matrix

$$M_1 = \begin{bmatrix} 1 & 0 & 0 \\ m_{21} & 1 & 0 \\ m_{31} & 0 & 1 \end{bmatrix}$$

where, in this case, $m_{21} = -1/3$, $m_{31} = -2/3$ are the multiplier. Then, the result is

$$M_1A = \begin{bmatrix} 3 & 1 & 6 \\ 0 & 2/3 & -1 \\ 0 & 1/3 & 1 \end{bmatrix} . \qquad (2)$$

Then, in the next step we apply a matrix

$$M_2 = \begin{bmatrix} 1 & 0 & 0 \\ 0 & 1 & 0 \\ 0 & m_{32} & 1 \end{bmatrix} ,$$

where $m_{32} = -1/2$. This has the effect of adding $m_{32} = -1/2$ times row 2 to row 3. The final result is

$$U = M_2M_1A = \begin{bmatrix} 3 & 1 & 6 \\ 0 & 2/3 & -1 \\ 0 & 0 & 3/2 \end{bmatrix} . \qquad (3)$$

We note that M_1, M_2 are nonsingular ($\det M_i = 1$), so we may multiply both sides by $M_1^{-1} M_2^{-1}$ to obtain

$$A = (M_1^{-1}M_2^{-1})U = LU,$$

where $L = M_1^{-1} M_2^{-1}$.

By comparing (1) with (2) and (3), note that the action of matrix M_1 is to set to zero all the subdiagonal elements of column 1 of A; then M_2 is used to set to zero all the subdiagonal elements of column 2.

In the general case, where A is $n \times n$, we must apply n-1 "M" matrices, one for each column. Matrix M_k, $k = 1,2,.., n-1$, has the following structure:

$$M_k = \begin{bmatrix} 1 & & & & & & \\ & 1 & & & & \bigcirc & \\ & & \ddots & & & & \\ & & & 1 & & & \\ & \bigcirc & & m_{k+1,k} & \ddots & & \\ & & & \vdots & & & \\ & & & m_{n,k} & & \bigcirc & 1 \end{bmatrix} .$$

↑

k-th column

Coefficients $m_{j,k}$, $j = k+1, k+2, \ldots, n$, will be referred to as the matrix multipliers.

The last item we need to complete the description of the LU decomposition is: what is "L"? To see what form has L, we note that the inverse of M_k, as can be easily verified, is the same as M_k with the multipliers in the k-th column negated:

$$M_k^{-1} = \begin{bmatrix} 1 & & & & & & \\ & 1 & & & & \bigcirc & \\ & & \ddots & & & & \\ & & & 1 & & & \\ & \bigcirc & & -m_{k+1,k} & \ddots & & \\ & & & \vdots & & & \\ & & & -m_{n,k} & & \bigcirc & 1 \end{bmatrix} .$$

↑

k-th column

Secondly, we note that when we form the product $L = M_1^{-1} \ldots M_{n-1}^{-1}$, the result is simply to fill in <u>all</u> the multipliers below the diagonal. In our 3×3 example, we have:

$$L = M_1^{-1}M_2^{-1} = \begin{bmatrix} 1 & 0 & 0 \\ 1/3 & 1 & 0 \\ 2/3 & 0 & 1 \end{bmatrix} \begin{bmatrix} 1 & 0 & 0 \\ 0 & 1 & 0 \\ 0 & 1/2 & 1 \end{bmatrix} = \begin{bmatrix} 1 & 0 & 0 \\ 1/3 & 1 & 0 \\ 2/3 & 1/2 & 1 \end{bmatrix} .$$

Here, one can see that the net effect is to collect all the multipliers from all the M_k, change sign and place them below the diagonal in their corresponding position. On the diagonal, we have all "1" 's. Hence, the (i,j) position of L, $i > j$, is $-m_{ij}$ = the multiplier used on row j when added to row i during the stage when column j is being zeroed:

$$L = \begin{bmatrix} 1 & & \bigcirc \\ & 1 & \\ (-m_{ij}) & & \ddots & \\ & & & 1 \end{bmatrix} .$$

In conclusion, we have found a decomposition for $A : A = LU$, where L is lower triangular with "1" 's on the diagonal and U is upper triangular.

What can we do with this? We give 2 uses of this decomposition:

A. Solve Linear Equations

By using $A = LU$, the system $Ax = b$ is equivalent to

$$LUx = b. \tag{4}$$

If we call

$$y = Ux,$$

we then reduce equation (4) to two triangular systems:

$$Ly = b$$

$$Ux = y.$$

Triangular systems are easily solved using the process known as "back-substitution". We note that if we apply the row operation of Gaussian Elimination also to the vector b, the result will be

$$M_{n-1}M_{n-2} \dots M_1 b = L^{-1}b = L^{-1}Ax = Ux = y.$$

Then the solution x can be found by solving

$$Ux = y.$$

It turns out that the extra work involved in applying the row operations also to b to obtain y is <u>exactly</u> the same as the work to solve $Ly = b$ for y. The two schemes are exactly equivalent, except that by using $Ly = b$, we may solve directly $Ax = \tilde{b}$, with a new right hand side $\tilde{b}$, without repeating the decomposition.

B. Computing Determinant

Since the determinant of a product of matrices is equal to the product of the determinants, we may immediately write the determinant of A as:

$$\det A = \det L \cdot \det U = 1 \cdot (u_{11}u_{22} \dots u_{nn}),$$

where u_{ij} is the (i,j) element of U. Here, we have used the well known fact that the determinant of a triangular matrix is simply the product of the diagonal elements.

Using the LU decomposition is <u>much</u> faster than using the definition of det A (Stewart, 1973).

2.2 - Orthogonal Decomposition

2.2.1 - QR *Decomposition*

In the LU decomposition, we have applied matrices that are not orthogonal; they do not preserve lengths or angles of vectors. Since 2 vectors may be made almost parallel by such transformations, we would like to see what one can do with orthogonal transformations, which do preserve lengths and angles.

A $n \times n$ matrix Q is orthogonal if its columns q_i are mutually ortho-normal, i.e.

$$q_i'q_j = \begin{cases} 0 & , \text{if } i \neq j \\ 1 & , \text{if } i = j, \end{cases}$$

or, in matrix notation,

$$Q'Q = I.$$

In this section, we show how one may triangularize a matrix using only orthogonal transformations, thereby preserving lengths and angles. Then we show why the use of orthogonal decompositions is particularly useful by giving an example of its use.

Consider the matrix:

$$A = \begin{bmatrix} 3 & 1 & 6 \\ 1 & 1 & 1 \\ 2 & 1 & 5 \end{bmatrix}$$

We would like to reduce the first column $[3 \quad 1 \quad 2]'$ to $[? \quad 0 \quad 0]'$, where "?" denotes a nonzero element whose value is to be determined. We see how to do this using a "rotation", i.e. a transformation of the form

$$Q_1 = \begin{bmatrix} c & s & 0 \\ -s & c & 0 \\ 0 & 0 & 1 \end{bmatrix}, \tag{5}$$

where, by orthogonality, $c^2 + s^2 = 1$.

We will use Q_1 to zero element a_{21}, i.e.

$$\begin{bmatrix} c & s & 0 \\ -s & c & 0 \\ 0 & 0 & 1 \end{bmatrix} \begin{bmatrix} 3 \\ 1 \\ 2 \end{bmatrix} = \begin{bmatrix} ? \\ 0 \\ 2 \end{bmatrix} .$$

The second line yields:

$$-s \cdot 3 + c \cdot 1 = 0,$$

which, together with $c^2 + s^2 = 1$, yields

$$c = 3/\sqrt{10} \quad , \quad s = 1/\sqrt{10} .$$

Having defined Q_1, we apply it to A to obtain Q_1A. Then, in the same way, we find the elements c and s of the rotation

$$Q_2 = \begin{bmatrix} c & 0 & s \\ 0 & 1 & 0 \\ -s & 0 & c \end{bmatrix} , \qquad (6)$$

to zero a_{31}. To complete the triangular decomposition, we need a third rotation Q_3 to zero a_{32}, obtaining finally

$$R = Q_3Q_2Q_1A.$$

In general, in the $n \times n$ case, we need $r = n(n+1)/2$ rotations to zero all the subdiagonal elements of A:

$$R = Q_rQ_{r-1} \cdots Q_1A = \text{an upper triangular matrix.}$$

Let

$$Q = (Q_rQ_{r-1} \cdots Q_1)^{-1}.$$

By orthogonality:

$$Q = Q_1' Q_2' \dots Q_r' \quad .$$

We have now the so called QR decomposition or orthogonal triangularization of A:

$$\underset{\text{arbitrary}}{\underset{\updownarrow}{A}} \;=\; \underset{\text{ortho}}{\underset{\updownarrow}{Q}} \;\cdot\; \underset{\text{upper triangular}}{\underset{\updownarrow}{R}}$$

2.2.2 - *Geometric Interpretation of a Rotation*

A rotation only affects 2 components of a vector, as can be seen from (5) and (6). Hence, we may look at a representative problem in the 2-space.

Consider the 2×2 rotation:

$$Q = \begin{bmatrix} c & s \\ -s & c \end{bmatrix} \qquad , \; c^2 + s^2 = 1.$$

We may represent $c = \cos\phi$, $s = \sin\phi$ for some angle ϕ .

Consider also a vector x of R^2, and denote by θ the angle between x and $e_1 = [1 \quad 0]'$:

$$x = \begin{bmatrix} x_1 \\ x_2 \end{bmatrix} = \|x\| \begin{bmatrix} \cos\theta \\ \sin\theta \end{bmatrix} .$$

Hence,

$$Qx = \|x\| \begin{bmatrix} c\cos\theta + s\sin\theta \\ -s\cos\theta + c\sin\theta \end{bmatrix} = \|x\| \begin{bmatrix} \cos(\theta-\phi) \\ \sin(\theta-\phi) \end{bmatrix} .$$

This means that Qx is the vector x rotated by angle ϕ.

2.2.3 QR *Decomposition by Householder Transformations*

As we have seen in 2.2.1, matrix Q of the QR decomposition can be obtained by multiplying $n(n+1)/2$ rotations. There is an alternative way of computing Q, based on the so-called "Householder transformations". The main advantage of these transformations is that one can zero out as many components of a vector as one likes by means of a single transformation, whereas, in general, by a rotation it is possible to zero out one component only. This implies that, to obtain the QR decomposition, one can use $n-1$ Householder transformations, instead of $n(n+1)/2$ rotations. The Householder transformation (or "elementary reflection") can be introduced by making reference to R^2 as follows. We want to transform a vector x to a vector v along e.g., axis

$$e_1 = [1 \quad 0] ,$$

such that $\| v \| = \| x \|$. This can be achieved by reflecting the vector around $x + v$ (see Fig. 1).

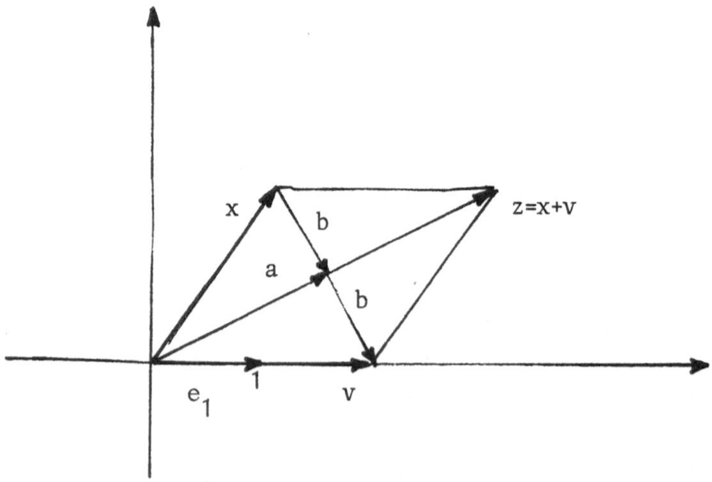

Figure 1.

We go through the following steps (note that we know x and $v = \|x\| e_1$).

- Axis of reflection $z = x + v = [x_1 + \|x\|, x_2]'$.
  The corresponding unit vector is $z/\|z\|$.

- Project x onto the axis of reflection to obtain:

$$a = \left[\frac{z}{\|z\|}\right]\left[\left(\frac{z}{\|z\|}\right)' x\right] = \frac{zz'x}{\|z\|^2}$$

- Find the difference between x and its projection:

  $b = a - x.$

- Reflect x around its projection a (or equivalently z):

$$v = x + 2b = x + 2(a-x) = 2a - x = -x + 2\frac{zz'x}{\|z\|} = -\left(I - 2\frac{zz'}{\|z\|^2}\right)x.$$

The matrix

$$P = I - 2\frac{zz'}{\|z\|^2},$$

gives the Householder transformation. Since

$v = -Px,$

we can conclude that, by such a linear transformation, one can "reflect" a vector x into any axis of the space.

In n-space, we can pick the desired target direction c so as to zero out at once as many components of a vector as we like.

2.2.4 *Solving Least Squares Problems Using Orthogonal Decompositions*

Let $A \in R^{m\times n}$, $m \geq n$, $b \in R^m$ and $x \in R^n$. The Linear Least Squares problem is the problem of finding the following minimum

$$\min_x \|Ax - b\| .$$

The algorithm of 2.2.1 may be applied to rectangular matrices just as well as square ones. In this case we see that the QR decomposition of the rectangular matrix A is:

$$A = Q \begin{bmatrix} R \\ O \end{bmatrix}, \qquad (8)$$

where $Q \in R^{m\times m}$ is orthogonal, $R \in R^{n\times n}$ is upper triangular and $O \in R^{(m-n)\times n}$ is a block of zero elements.

As Q is orthogonal, it does not change the norm. Hence:

$$\|Ax - b\| = \|Q' (Ax - b)\| = \| \begin{bmatrix} R \\ O \end{bmatrix} x - c \|,$$

where

$$c = Q'b.$$

Partitioning this vector conformally with $\begin{bmatrix} R \\ O \end{bmatrix}$,

$$c = \begin{bmatrix} c_1 \\ c_2 \end{bmatrix},$$

we have

$$\| Ax - b \| = \| \begin{bmatrix} Rx - c_1 \\ c_2 \end{bmatrix} \| .$$

In order to minimize this norm, we set

$$x = R^{-1}c_1. \tag{9}$$

Thus,

$$\min_x \| Ax - b \| = \| c_2 \|.$$

To find the optimum value of x given by (9) we have to solve system

$$Rx = c_1. \tag{10}$$

In this respect, it is worth noticing that a direct minimization of $\| Ax - b \|$ leads to the celebrated normal equations:

$$A'Ax = A'b. \tag{11}$$

In view of (8), system (11) is equivalent to

$$R'Rx = R'c_1. \tag{12}$$

It is important to observe that solving system (10) is preferable to solving (12) as one obtains fewer errors. For a complete discussion of this, one must use the error analysis of Linear Systems, see e.g. (Lawson and Hanson, 1974). However, we may give an example of the problem.

Suppose we are working on a computer with precision 10^{-7}, i.e. we carry only 7 significant digits. Consider matrix

$$A = \begin{bmatrix} 1 & 1 \\ 10^{-4} & 0 \\ 0 & 10^{-4} \end{bmatrix}. \tag{13}$$

A has rank 2, but if we form

$$A'A = \begin{bmatrix} 1 + 10^{-8} & 1 \\ 1 & 1 + 10^{-8} \end{bmatrix},$$

in our computer we will loose the part "10^{-8}" and obtain instead

$$A'A = \begin{bmatrix} 1 & 1 \\ 1 & 1 \end{bmatrix},$$

which has rank 1. So, we loose rank information.

3. Special Forms Used in Numerical Linear Algebra-Why

The LU and QR decompositions discussed above are used as the basic step in the computation of the decompositions to be introduced in the following. The previous section also serves to give the flavour of the approach used in the rest of this work.
We now concentrate on the problems of finding a number of useful things about a given square matrix A, i.e.

(i) Eigenvalues: λ_i

(ii) Determinant: det A

(iii) Rank of A

(iv) Nullspace of A: ker A

(v) Image or Range of A: colsp A.

3.1 - The Jordan Canonical Form

As is well known, there are many classical decompositions for matrices, the most common being the Jordan decomposition. Matrix A is then decomposed into the product of 3 matrices:

$$A = PJP^{-1},$$

where P is nonsingular, and J is in the so-called Jordan Canonical Form (Gantmacher, 1959). This form can tell us the eigenvalues of A (elements of the diagonal of J), the determinant (product of the diagonal elements of J) and the rank (dimension of the matrix minus the number of Jordan bloks corresponding to $\lambda_i = 0$). Furthermore, the columns of P corresponding to all zero columns of J generate the nullspace of A, whereas the columns corresponding to nonzero rows of J generate the range of A.

So, the Jordan Canonical Form enables one to find out all items (i) - (v). However, for numerical computations, this form is ill-advised. The matrix P can be extremely ill-conditioned (i.e. almost singular), especially if the eigenvalues λ_i are poorly separated(almost coinciding).But even if the λ_i are "well separated", finding the Jordan Form is an ill-posed problem. This calls for the question of

3.2 - Numerical Conditioning of a Problem

Numerical conditioning is an important consideration whenever one considers the use of a digital computer. Because of the finite word length, numbers can be represented only approximately in the computer. In the treatment of such approximations, the model most often used is to consider what happens if the numbers are perturbed slightly.

The eigenvalues can be extremely sensitive to perturbations: take for example the 3×3 matrix

$$A = \begin{bmatrix} -64 & 82 & 21 \\ 144 & -178 & -46 \\ -778 & 962 & 248 \end{bmatrix} , \tag{14}$$

which has eigenvalues 1, 2, 3. If we add a small perturbation εE, where

$$E = 10^{-4} \begin{bmatrix} 0.5 & -0.1 & 5.1 \\ -0.6 & 1.1 & -6.2 \\ -0.1 & 0.3 & -1.6 \end{bmatrix}$$

is a rank one matrix of norm $\sim 10^{-3}$, then, for any $\varepsilon > 0.45$, the perturbed matrix $A + \varepsilon E$ will have complex eigenvalues! This shows that problems may occur even on small innocuous-looking matrices!

Even with 7-16 digits of accuracy, this is a relevant problem: In the following 20×20 examples, due to Wilkinson (1965), perturbations in the 9th place in some element will completely destroy <u>all</u> digits of accuracy in the eigenvalues- even the order of magnitude will be wrong in some cases:

$$A = \begin{bmatrix}
20 & & & & & & & & & & \\
20 & 20 & & & & & & & & & \\
 & 19 & 20 & & & & \text{O} & & & & \\
 & & 18 & 20 & & & & & & & \\
 & & & 17 & 20 & & & & & & \\
 & & & & 16 & 20 & & & & & \\
 & & & & & \ddots & \ddots & & & & \\
 & & & & & & 5 & 20 & & & \\
 & & \text{O} & & & & & 4 & 20 & & \\
 & & & & & & & & 3 & 20 & \\
 & & & & & & & & & 2 & 20 \\
 & & & & & & & & & & 1
\end{bmatrix} \tag{15}$$

We point out that to obtain the solution to a problem one must apply an algorithm. If a problem is ill-posed or ill-conditioned then slight perturbations to the data (such as that occuring in the computer from the finite word length) can destroy the desired result, as was seen in example (15). In this case there is no algorithm that can give satisfactory results.

In case a problem is not so badly conditioned, then we may consider methods for its solution. In this regard we would like to use methods which introduce as few errors as possible, causing the smallest perturbations, or at least providing the best bounds on such errors. These desirable properties are called collectively the stability of an algorithm. If a problem is well-conditioned and the method to solve it is stable then we may believe the answer (to the limits defined by the conditioning of the problem and the stability of the algorithm). If the problem is well conditioned, but the method is unstable, then we can hope for a better algorithm; but in the opposite case, we cannot hope to improve the accuracy of a solution for a badly conditioned problem by any algorithm, stable or not.

The usual measure of stability for algorithms in Linear Algebra is so-called "Backward Stability". This is probably best introduced in terms of a specific algorithm in the next section.

4. Schur Decomposition

The matrix decomposition which is numerically useful and which most closely corresponds to the Jordan decomposition is the so-called Schur decomposition. Denoting by "*" conjugate transpose, this decomposition is given by

$$A = Q R Q^*, \tag{16}$$

where Q is a (possibly complex) orthogonal matrix ($Q^*Q = I$) and R is a (possibly complex) upper triangular matrix.

We first note the mathematical relevance of the Schur decomposition: since (16) is a similarity transformation, the eigenvalues of A are those of R, for which the eigenvalues are just the diagonal elements. The determinants satisfy the relation:

$$\det A = \det Q \det R \det Q^* = \det R =$$
$$= \text{product of the diagonal elements of } R.$$

Hence, the Schur form yields items (i) and (ii) of Sect.3.

What about the numerical properties? Since Q is orthogonal, it is always bounded in size and well-conditioned (the columns are never "almost parallel", so that Q is never "close to singularity").

Though no algorithm can hope to remove the ill-conditioning in the original problem, we can make sure that the result is not more sensitive to perturbations than the starting problem. This is the distinction between <u>ill-conditioning in a problem</u> and <u>instability of an algorithm</u> mentioned in Sect. 3.2. The advantage of the Schur form (16) is that perturbations in A will result in perturbations of the <u>same size</u> in R, so the sensitivity to perturbations is <u>not made worse</u>. In general, any algorithm/form based on unitary transformations can be shown (Wilkinson, 1965) to be "backward" stable, that is the result is exact for a problem close to the original one.

Instead of saying (a) the algorithm is <u>forward stable</u>: "The algorithm has produced an answer $\tilde{R}$ close to the R that we would hope to obtain with exact arithmetic starting with the exact original problem A", we say (b) the algorithm is <u>backward stable</u>: "The algorithm has produced an answer $\tilde{R}$ which is <u>exact</u> for a slightly perturbed starting problem $\tilde{A}$ close to the original problem A". Here, "close" means on the order of the precision of the computer used. As is seen from examples (14) and (15), the statement (a) is not true in general-slight changes to A can complete destroy R.

By using orthogonal transformations, we may still obtain the exact eigenvalues for a matrix close to A (backward stability).

In exact arithmetic, we make no errors, hence we obtain exact eigenvalues of (14) and (15). If our method makes slight errors no method can satisfy (a), but we can still hope for (b). These examples shows the best one can expect in a method. Unfortunately, the Schur form cannot be used reliably for items (iii)-(v). This can be illustrated as follows. If A is already upper triangular, the Schur form is obtained with R = A, Q = I. Consider then the n × n matrix:

$$A = \begin{bmatrix} 1 & -1 & & -1 \\ & 1 & & \\ & & \ddots & -1 \\ \bigcirc & & & 1 \end{bmatrix} + \alpha \begin{bmatrix} & & \bigcirc \\ & & \\ 1 & & \end{bmatrix} \qquad (17)$$

If $\alpha = 0$, all the eigenvalues are exactly 1. Nevertheless, the rank is close to deficient, since "perturbing" α to $\alpha = -1/2^{(n-2)}$ will make A singular! This can be verified by forming Ax with

$$x = \left[\alpha^{-1}, \alpha^{-2}, \alpha^{-3}, \ldots, \alpha^{-(n-\alpha)}, \alpha^{-(n-1)}, \alpha^{-(n-1}\right]$$

Therefore, we need a better way to find the rank, det, etc. of a matrix. Such a way is provided by the Singular Value Decomposition.

5. Singular Value Decomposition-Condition Number of a Matrix

In this section, we introduce another decomposition relevant to items such as rank: the Singular Value Decomposition (S.V.D.). We also introduce the concept of Condition Number of a matrix and try to explain its significance.

The S.V.D. of a $m \times p$ matrix A is

$$A = U \Sigma V^* \tag{18}$$

where U and V are square and orthogonal matrices and Σ is a $m \times p$ real and diagonal matrix, with nonnegative diagonal elements. By letting $n = \min(m,p)$, we usually assume that

$$\Sigma = \mathrm{diag}\,(\sigma_1, \sigma_2, \ldots, \sigma_n),$$

$$\sigma_1 \geq \sigma_2 \geq \ \ldots \geq \sigma_n \geq 0.$$

In what follows, we will use the matrix 2-norm:

$$\|A\| = \max_{\|x\| = 1} \|Ax\| , \tag{19}$$

where $\|x\|$ is the usual vector 2-norm. In this norm, we have several properties. The most immediately relevant property is the fact that orthogonal matrices do not affect (Stewart, 1973) the 2-norm of a matrix or vector:

$$\|Qx\| = \|x\| \tag{20a}$$

$$\|QA\| = \|A\|. \tag{20b}$$

Notice also that $\|Q\| = 1$.

What kind of information can one gleam from the S.V.D.? The most obvious is the norm of A. From (18), (20) and (19):

$$\|A\| = \|\Sigma\| = \sigma_1.$$

If A is a nonsingular square matrix, its inverse is given by

$$A^{-1} = V \Sigma^{-1} U^*.$$

Moreover,

$$||A^{-1}|| = || \Sigma^{-1}|| = \sigma_n^{-1}.$$

Given a square nonsingular matrix A, the number

$$k(A) = || A || \cdot || A^{-1}|| \tag{21}$$

is said to be the condition number of A. Obviously,

$$k(A) = \sigma_1/\sigma_n.$$

The condition number happens to be a very useful quantity in estimating the sensitivity of such items as rank, determinant, inverse, solution to a set of linear equations, etc., with respect to perturbations to the matrix A. It also gives the "distance to singularity".

To see this, we start with the classical origin of k(A). Consider the problem of solving the matrix equation $Ax = b$. When using a computer, we obtain an approximate result $\tilde{x}$, which we consider exact for the slightly perturbed problem $A\tilde{x} = \tilde{b}$. Note that we have perturbed only b, not A. We have:

$$Ax = b$$

$$A\tilde{x} = \tilde{b}.$$

Subtract to get

$$A(\tilde{x} - x) = \tilde{b} - b.$$

Multiplying both sides by A^{-1} and taking the norms, one obtains:

$$||\tilde{x} - x|| \leq || A^{-1}|| \; || \tilde{b} - b||, \tag{22}$$

i.e. the (error in answer) is bounded by the (error in right hand side) magnified by $||A^{-1}||$. However, to estimate the number of digits of accuracy in $\tilde{x}$, we need the "relative error"

$$\frac{\|\tilde{x} - x\|}{\|x\|} .$$

If the relative error is e.g. $\sim 10^{-6}$, then we have about 6 digits of accuracy, regardless of the size of x. To obtain an estimate of the relative error, we use the relation $Ax = b$ to obtain:

$$\|A\| \ \|x\| \geq \|b\|$$

i.e.

$$\frac{\|A\|}{\|b\|} \geq \frac{1}{\|x\|} . \tag{23}$$

From (22) and (23), and definition (21),

$$\frac{\|\tilde{x} - x\|}{\|x\|} \leq k(A) \frac{\|\tilde{b} - b\|}{\|b\|} \tag{24a}$$

i.e. the (number of good digits in x) is bounded by the (number of good digits in b) magnified by k(A).

For perturbations in A, one can obtain an analogous result:

$$\frac{\|\tilde{x} - x\|}{\|\tilde{x}\|} \leq k(A) \frac{\|\tilde{A} - A\|}{\|\tilde{A}\|} \tag{24b}$$

Here, the errors are relative to the approximate values.

We give a few examples of condition numbers of some particular matrices:

(a) $k(I) = 1$

(b) $k(A) \geq 1$, for any A.

Indeed:

$$k(A) = \|A\| \; \|A^{-1}\| \geq \|A A^{-1}\| = \|I\| = 1$$

(c) $k(Q) = 1$, if Q is orthogonal

(d) Let T_6 be the 6×6 Hilbert matrix, the elements of which are $t_{ij} = (1+i+j)^{-1}$.
Then,
$k(T_6) \sim 10^6$.

The S.V.D. and $k(A)$ can be used to find the "distance to singularity" of a matrix A. We note that

rank A = rank Σ = number of nonzero σ_i's.

In particular, if A is nonsingular, $\sigma_i > 0$, $\forall i$. Now, suppose that A is nonsingular and E is a perturbation such that $A + E$ is singular. Then, we may write, using the S.V.D. of A (18):

$$V^* (A + E) U = V^* A U + V^* E U = \Sigma + F,$$

where

$$F = V^* E U.$$

Because U, V are orthogonal, $\| E \| = \| F \|$. Thus, perturbations E to A correspond exactly to perturbations F to Σ.

We can define "distance to singularity" as the norm of the smallest E such that $A + E$ is singular:

$$d_{sing} = \min_{A+E \text{ sing.}} \| E \| . \tag{25}$$

In view of the discussion above, this corresponds to

$$d_{sing} = \min_{\Sigma+F \text{ sing.}} \| F \| . \tag{26}$$

Since $\Sigma = \mathrm{diag}(\sigma_1, \sigma_2, \ldots, \sigma_n)$, with $\sigma_1 \geq \sigma_2 \geq \ldots \geq \sigma_n > 0$, it is clear that the F which achieves the minimum in (26) is

$$\tilde{F} = \mathrm{diag}(0,0,\ldots,0,-\sigma_n) \; ,$$

so that

$$\| \tilde{F} \| = \sigma_n .$$

Hence, the E achieving the minimum in (25) is

$$\tilde{E} = U \tilde{F} V^* = - \sigma_n u_n v_n^* \; ,$$

where we have labeled the columns of U,V:

$$U = \left[u_1 \; u_2 \; \cdots \; u_n \right]$$

$$V = \left[v_1 \; v_2 \; \cdots \; v_n \right] .$$

Notice also that

$$\| \tilde{E} \| = \sigma_n .$$

Hence, the distance to singularity is

$$d_{sing} = \sigma_n . \tag{27}$$

Consequently, the "relative distance to singularity" (relative to the size of the starting matrix A) is:

$$\frac{d_{sing}}{\| A \|} = \frac{\sigma_n}{\sigma_1} = k(A) . \tag{28}$$

So, k(A) not only indicates the difficulty one can expect in solving Ax = b, but also shows how close A is to a singular matrix relative to the size of the matrix. In other words, k(A) gives the sensitivity of rank A to perturbations in A.

It should be noted that the quantity $k(A) = \|A\| \, \|A^{-1}\|$ can be defined using any matrix norm corresponding to a vector norm, and relations (24) hold in any such norm. The results involving the S.V.D. however, are valid only in the 2-norm.

We have seen how the S.V.D. can be used to obtain such quantities as $\|A\|$, rank A, k(A) . What about points (iv) and (v) of Sect.3, spaces ker A, colsp A? For this case, we start with a singular $n \times n$ A. Let us suppose that the singular values of A satisfy:

$$\sigma_1 \geq \sigma_2 \geq \dots \geq \sigma_r > \sigma_{r+1} = \sigma_{r+2} = \dots = \sigma_n = 0, \text{ so that rank } A = r.$$

We note in passing that, analogous to (27) and (28), the quantity σ_r gives the size of the smallest perturbation needed to further reduce rank A, and (σ_r/σ_1) gives the relative size of such perturbation.

We write the S.V.D. of A as

$$A = U \begin{bmatrix} \sigma_1 & & & & & & \\ & \sigma_2 & & & & O & \\ & & \ddots & & & & \\ & & & \sigma_r & & & \\ & & & & 0 & & \\ & O & & & & 0 & \\ & & & & & & \ddots \\ & & & & & & & 0 \end{bmatrix} V^* = \begin{bmatrix} U_1 & U_2 \end{bmatrix} \begin{bmatrix} \Sigma_1 & 0 \\ 0 & 0 \end{bmatrix} \begin{bmatrix} V_1^* \\ V_2^* \end{bmatrix} \qquad (29)$$

where

$$\Sigma_1 = \operatorname{diag}(\sigma_1, \sigma_2, \dots, \sigma_r)$$

is $r \times r$ and nonsingular. U, V have been partitioned conformally to Σ. Thus,

$$A = U_1 \Sigma_1 V_1^* ,$$

where U_1, V_1 are $n \times r$ orthonormal matrices and Σ_1 is $r \times r$ nonsingular. Hence, U_1 is an orthonormal basis for colsp A, and ker A is the orthogonal complement of the space with V_1 as orthonormal basis i.e. V_2 is an orthonormal basis for ker A.

In practice, to use (29), one will frequently encounter the situation

$$\sigma_1 \geq \sigma_2 \geq \ldots \geq \sigma_r > \sigma_{r+1} \geq \ldots \geq \sigma_n \geq 0,$$

where some of the later singular values are "small", i.e. on the order of e.g. the machine precision. The problem is to decide "how small is small", that is at what point to consider a small singular value zero.

It is best to illustrate the problem here. Assume that the singular values are (scaled so that $\sigma_1 = 1$) e.g.

$$10^0 \quad 10^{-1} \quad 10^{-1} \quad 10^{-2} \quad 10^{-4} \quad 10^{-8} \quad 10^{-9} \quad 10^{-10} \quad 0 \quad 0,$$

where only the order of magnitude are shown. Then we see that the exact rank of the matrix is 8. But if the original data only had 6 digits of accuracy, then any number $< 10^{-6}$ should be considered zero. Hence we would consider the rank to be 5.

If, instead, we had the values

$$10^0 \quad 10^{-2} \quad 10^{-4} \quad 10^{-6} \quad 10^{-8} \quad 10^{-10} \quad 10^{-12} \quad 10^{-14} \quad 10^{-16} \quad 10^{-18}$$

or

$$10^0 \quad 10^{-1} \quad 10^{-2} \quad 10^{-3} \quad 10^{-4} \quad 10^{-5} \quad 10^{-6} \quad 10^{-7} \quad 10^{-8} \quad 10^{-9},$$

then there is no obvious gap, so the effective rank depends almost entirely on the choice of the zero tolerance!

Unfortunately, this situation can and does frequently arise in practice, especially in large matrices. This is not a defect of the S.V.D. If this situation arises, it really means that a small (negligible) perturbation to A will reduce the rank, and another perturbation, only slightly larger, can reduce the rank even further. For a full discussion of the S.V.D. and the rank see (Klema and Laub, 1980).

We close this section with a few examples of situations involving the S.V.D. We just point out the idea, leaving the details to the reader.

a) Least Squares (Lawson & Hanson, 1974)

This was the classical origin of the S.V.D. It is useful to solve problem (7) in cases where A is rank deficient. In such cases, we cannot use the QR decomposition because the matrix R in (8) is singular, and hence we cannot solve (9).

If we resort instead to the S.V.D. of A, we have, in view of (20a)

$$\|Ax - b\| = \|U\Sigma V^* x - b\| = \|\Sigma V^* x - U^* b\| = \|\Sigma y - c\|,$$

where $y = V^* x$ and $c = U^* b$. The result is we have converted the original problem to a diagonal problem involving Σ. We partition the above as in (29) to obtain

$$\left\| \begin{bmatrix} \Sigma_1 & 0 \\ 0 & 0 \end{bmatrix} \begin{bmatrix} y_1 \\ y_2 \end{bmatrix} - \begin{bmatrix} c_1 \\ c_2 \end{bmatrix} \right\| .$$

We minimize this norm by setting $y_1 = \Sigma_1^{-1} c_1$. In the solution, we find that y_2 is free!

(b) Pseudo Inverse

The pseudo inverse A^+ of A, Lawson and Hanson, 1974, can be expressed as

$$A^+ = V \begin{bmatrix} \Sigma_1^{-1} & 0 \\ 0 & 0 \end{bmatrix} U^* ,$$

where we have used the partitioning (29).

(c) Relation to A^*A

We point out the relationship of the S.V.D. to the classical idea of eigenvalues.

If

$$A = U \Sigma V^*,$$

then

$$A^*A = V \Sigma U^* U \Sigma V^* = V \Sigma^2 V^* ,$$

$$\Sigma^2 = \text{diag}(\sigma_1^2, \sigma_2^2 , \ldots , \sigma_n^2).$$

Hence, the singular values σ_i are just the square root of the eigenvalues of A^*A, and the columns of V are the eigenvectors of A^*A. In fact, using the fact that A^*A is symmetric positive semi-definite for any A, one can carry this argument in reverse to prove the existence of the S.V.D. (Lawson and Hanson, 1974).

However, for actual computation of the S.V.D. or the solution of the least squares problem (a), it is almost always more accurate and faster to compute the S.V.D. directly without forming A^*A, as can be seen from example (13). In the case of 2×2 matrix problems (or $2 \times n$ for any n), the results obtained using A^*A are often sufficiently accurate, especially when performing computation by hand.

6. Applications of Previous to Linear Systems

We finally take a look of how the previous discussion on numerical stability applies to Linear Dynamic Systems. Consider either the continuous-time system

$$\begin{aligned} \dot{x}(t) &= F\,x(t) + G\,u(t) \\ y(t) &= H\,x(t) \end{aligned} \tag{30}$$

or the discrete-time system

$$\begin{aligned} x(t+1) &= F\,x(t) + G\,u(t) \\ y(t) &= H\,x(t) \end{aligned} \tag{31}$$

where F is $n \times n$, G is $n \times m$ and H is $p \times n$.

The so-called Markov parameters are defined as

$$w(i) = H\,F^{i-1}\,G \qquad , \; i = 1,2,\ldots$$

In discrete time they are the values of the pulse response whereas in continuous-time they are the values of the pulse response at the origin and its derivatives.

We shall consider as particular examples two problems from the point of view of numerical stability. (i) Determining whether a given system is reachable and (ii) Determining the system order n from the external sequence $w(\cdot)$.

As for problem (i), we recall that (Kalman, Falb, Arbib, 1969) a system is said to be reachable if, for each state $\bar{x}$ and each time point t, there exist a $\tau<t$ and an input function $u(\cdot)$ defined over $[\tau,t)$ such as to carry $x(\tau) = 0$ into $x(t) = \bar{x}$. Obviously, reachability depends upon matrices F, G only.

The problem of determining whether a given system is reachable or not has been studied extensively in the last 30 years, and many criteria for reachability have been developed. Examples of popular criteria for continuous-time systems are given in this theorem:

Theorem A

The system (30) is reachable if and only if any of the following equivalent conditions is true

(C1) The $n \times (nm)$ matrix $\mathcal{C} = [G\ FG\ F^2G \ldots F^{n-1}G]$ has rank n (full rank)

(C2) The $n \times (nm)$ matrix $P(s) = [sI-F \,\vdots\, G]$ has rank n for any complex number s (PBH test).

(C3) There exists a $m \times n$ matrix K such that the poles (eigenvalues) of $F + GK$ are all different from those of F (state feedback).

(C4) (In case F is stable, i.e. all the eigenvalues of F have negative real part). The grammian W has rank n, where W is the solution of the Lyapunov equation:

$$FW + WF = -GG' \tag{32}$$

(Grammian condition)

(C5) There exists no pair $(\tilde{F},\tilde{G})$ related to (30) by $\tilde{F} = TFT^{-1}$, $\tilde{G} = TG$, where T is some nonsingular matrix, such that $\tilde{F}$, $\tilde{G}$ can be partitioned conformally as

$$\tilde{F} = \begin{bmatrix} \tilde{F}_{11} & \tilde{F}_{12} \\ 0 & \tilde{F}_{22} \end{bmatrix}, \qquad \tilde{G} = \begin{bmatrix} \tilde{G}_1 \\ 0 \end{bmatrix}. \tag{33}$$

In (33), $\tilde{F}_{11}$ and $\tilde{F}_{22}$ are square matrices.

This condition can be read as: "If there is a pair $(\tilde{F},\tilde{G})$ of the form (33), then the system is not reachable". □

For discrete-time system, we have an analogous theorem:

Theorem B

The system (31) is reachable if and only if any of the equivalent conditions (C1), (C2), (C3) or (C5) holds, or, equivalently, if and only if the discrete-time Grammian W: = $\mathcal{C}\,\mathcal{C}'$ is non-singular. □

We discuss this theorem from a numerical point of view by first showing the limitations of (C1), (C2), (C3), (C4) and then giving some positive results for conditions (C5), (C1), (C3).

Many of the conditions in Theorems A and B depend on the computation of the rank of a matrix or sub-matrix. This can lead to problems when the rank problem becomes ill-posed. Frequently, a submatrix will be rank-deficient, not exactly but only to the working precision of our computer. In this case we might conclude that the starting system is "unreachable to working precision", or "almost unreachable". For example, regarding criterion (C1), if we take the example of Paige (1981):

$$\begin{aligned} F &= \mathrm{diag}\,(1,\ 1/2,\ 1/2^2,\ldots,1/2^{n-1}) \\ G &= [1,\ 1,\ 1,\ldots,1]' \end{aligned} \qquad (34)$$

then by inspection it is apparent that the system is reachable. However if we form the matrix $\mathcal{C}$ for n=10, we find it has a smallest singular value $\sigma_{min} = 10^{-12}$, so we would conclude that $\mathcal{C}$ has rank $\leq n-1$, i.e. the system defined by the pair (34) is not reachable. The exact rank of $\mathcal{C}$ is n (all singular values are non-zero), but, to the working precision of most computers, 10^{-12} is considered zero.

A similar problem arises when considering the problem of determining the system order n from the external sequence w(·).

This problem is indeed the problem of finding the rank of a suitable matrix. Precisely, let

$$\mathcal{H}_r = \begin{bmatrix} w(1) & w(2) & \dots & w(r) \\ w(2) & w(3) & \dots & w(r+1) \\ \cdot & \cdot & \dots & \cdot \\ w(r) & w(r+1) & \dots & w(2r-1) \end{bmatrix}$$

be the so-called r-dimension Hankel matrix.

We recall that, given a system defined by the triple (F,G,H), the system defined by (F', H', G') is named the dual system and that a system is said to be observable if its dual is reachable (see Kalman, Falb, Arbib, 1961). Then, from (C1) of Theorems A and B it follows that the system is observable if and only if the matrix

$$\mathcal{O} = \left[H' \quad F'H' \quad (F')^2H' \quad \dots\dots \quad (F')^{n-1}\, H' \right]'$$

is full rank.

It is easy to see that

$$\mathcal{H}_n = \mathcal{O}\ \mathcal{C} \ .$$

Therefore, if a system is both reachable and observable,

rank $\mathcal{H}_n = n$.

In fact, it can be shown that a system is reachable and observable if and only if

rank $\mathcal{H}_r = n \quad , \qquad \forall\ r \geq n.$

This means that, in principle, the order of a reachable and observable system can be found from the pulse response, by determining the rank of a Hankel matrix of large dimension.

Now suppose that F and G are given by (34), and H = G'. In this case the controllability and observability matrices are symmetric and equal : $\mathcal{C} = \mathcal{C}' = \mathcal{O}$. Hence $\mathcal{H}_n = \mathcal{C}^2$ and $\sigma_{min}\mathcal{H}_n = (\sigma_{min}\mathcal{C})^2$. We give in Table 1 the singular values of $\mathcal{C}$ and $\mathcal{H}_n$ for various values of n.

n	$\sigma_{min}\mathcal{C}$	$\sigma_{min}\mathcal{H}_n$
5	8.8×10^{-4}	7.7×10^{-7}
6	4.8×10^{-5}	2.3×10^{-9}
7	1.4×10^{-6}	1.9×10^{-12}
8	2.1×10^{-8}	4.3×10^{-16}

Table 1

Depending on the precision of the computer being used, matrix $\mathcal{H}_n$ could be considered rank-deficient even though the matrix $\mathcal{C}$ is not, and hence the system is reachable.

The discussion above motivates the introduction of the concept of "almost reachability". With this objective in mind, we formally define the concept of distance to the nearest unreachable system.

<u>Definition</u> (Paige, 1981), (Eising, 1984)

Given a system (30) reachable, we say it has a distance $\mu(F,G)$ from an unreachable system if

(a) The pair $(\bar{F}, \bar{G}) := (F + \delta F,\ G + \delta G)$ is not reachable, with $\|\delta F \vdots \delta G\| = \mu$

and

(b) For any pair $(\bar{\bar{F}}, \bar{\bar{G}})$ with $\| \bar{\bar{F}} - \bar{F} \vdots \bar{\bar{G}} - G \| < \mu$ the pair $(\bar{\bar{F}}, \bar{\bar{G}})$ is reachable. □

In other words, μ is the norm of the smallest perturbation δF, δG to F and G of (30) that yields an unreachable system.

Miminis (1981) has found a more computational way to define μ in terms of (C2):

$$\mu = \min_{s \in C} \sigma_n (P(s)) \tag{35}$$

where σ_n denotes the n-th (smallest) singular value of P(s). We take the minimum over all complex numbers; in fact in some cases the s* achieving the minimum is not real, as illustrated by the example

$$F = \begin{bmatrix} 0 & -1 \\ 1 & 0 \end{bmatrix} \qquad G = \begin{bmatrix} 1 \\ 0 \end{bmatrix} \tag{36}$$

for which it has been shown, Boley and Lu (1984) that $\mu = .6614$ achieved when $s = \pm i \frac{\sqrt{15}}{4}$, and that the minimum is not achieved for any real s. Hence, in this case, the perturbation δF, δG is not real.

Paige (1981) points out the numerical problems one may encounter using (C1) - (C3). An example for (C1) we already mentioned. For (C2), one may show that, if rank $P(s) < n$, then s must be an eigenvalue of F. One then has a choice of limiting one search to just eigenvalues of F, or doing it over the whole complex plane. The problem with the former is the fact that in some cases, one may be unable to compute the eigenvalues of F to any accuracy (e.g. if F is the matrix (15)) (Paige, 1981). In the latter case, the computation involved can become prohibitively expensive, Miminis (1981). Eising (1982) has also described a way to compute μ in terms of an n-dimensional minimization problem, which can also be expensive.

In the case of (C3), we can encounter again the same problem as in (C2) in that the success in distinguishing the poles depends critically on the eigenvalue problem, which may be badly conditioned((15) is an extreme example).

We point our that using the Grammian can lead to similar problems, because

(a) in the discrete time case, the grammian W is defined as $\mathcal{C}\mathcal{C}'$, which we have seen is a poor vehicle in this regard

(b) in the continuous time case the determination of reachability, depends on the solution of the Lyapunov equation (32), which, among other things, becomes ill-conditioned when F is <u>almost unstable</u>.

We can say a few positive things about (C1)-(C5). First of all, with regard to (C5), several people, Paige (1981), Van Dooren et al. (1979), have given an algorithm ("staircase algorithm") that finds a unitary transformation T exhibiting the form (33) if such exists. This algorithm is very similar to one used to compute the Jordan canonical form, Kublanovskaya (1961 and 1968). Unfortunately, if the system (30) is reachable, but almost unreachable, then there is no easy way to estimate the distance μ. In particular, the "staircase algorithm" does not give a good estimate for μ. In fact, the system (36) is already in the form produced by the "staircase algorithm". The best estimate one can obtain by inspection of (36) is $\mu \leq 1$ since a nearby unreachable system can be obtained by setting F_{21} to 0 in (36). There is no obvious way to see from (36) that μ is actually .6614, a value obtained in Boley and Lu (1984) by going back to first principles and using simple algebraic argument.

With regard to condition (C1) one can show that small singular values of $\mathcal{C}$ do not necessarily imply almost unreachability of the system (F,G), but under certain circumstances one can bound the distance μ in terms of the singular values of $\mathcal{C}$:

<u>Theorem</u> (Boley and Lu, 1984)

If $\mathcal{C}$ has singular values $\gamma_1 \geq \dots \geq \gamma_{n-1} \geq \gamma_n > 0$ and $\alpha_o + \alpha_1\lambda + \dots + \alpha_n\lambda^n$ is the characteristic polynomial of F, with $\alpha_n = 1$, then

$$\mu \leq \sqrt{nm} \quad (1 + \max |\alpha_i|) \frac{\gamma_n}{\gamma_{n-1}} \tag{37}$$

where F is $n \times n$, G $n \times m$.□

This theorem says that the "distance to unreachable (μ) depends not on the <u>size</u> of the smallest singular values (γ_n) but on the <u>spread</u> of the singular values (γ_n/γ_{n-1}). In the example, Van Dooren (1981):

$$F = \begin{bmatrix} -1/2 & 0 \\ \sqrt{\varepsilon} & -1/2 \end{bmatrix} \qquad G = \begin{bmatrix} \sqrt{\varepsilon} \\ 0 \end{bmatrix} \tag{38}$$

we find that $\mathcal{C}$ has two singular values on the order of $\sqrt{\varepsilon}$ and ε. One can show that the distance μ of (38) is not $O(\varepsilon)$, but $O(\sqrt{\varepsilon})$, (Van Dooren (1981)). Hence the criterion (C1) suitably modified, is still of some use. In fact Boley and Lu (1984) showed how to use (37) to obtain a nearby <u>real</u> unreachable system, something which may have more physical significance than a complex one (see e.g. discussion around eq. (36)).

Finally, with regard to (C3) we can also give a result that shows that in at least one direction, this condition can give useful results:

Theorem (Boley and Lu, 1984)

Assume that pair (A,B) is reachable and that λ_n is a simple eigenvalue of A. Then for any h > 0 less than some sufficiently small h_o, there exists a feedback matrix K with $||K|| \leq h$ such that all the eigenvalues $\nu_1,\dots,\nu_n$ of the closed loop matrix A + BK differ from λ_n by at least $h\,\mu(A,B)$.$\square$

What this theorem says is that if some eigenvalue of A is moved by any state feedback K no more than $\varepsilon||K||$, as $||K||$ asymptotically approaches zero, then $\mu \leq \varepsilon$. In other words, if some eigenvalue is hard to move by any small state feedback, then the system is almost unreachable.

In this direction this is a positive result, but as mentioned above the converse is not true, since the poles may move due to ill-conditioning of the eigenproblem rather than the reachability property.

We have given in this section just a few examples of the limitations of some of the classical criteria for reachability from a numerical point of view, and some of the positive aspects to counterbalance these limitations.

References

Boley, D.L. and W.S. Lu, 1984: The Quasi Kalman Decomposition and State Feedback. American Control Conference, S. Diego.

Gantmacher, F.,1959: Theory of Matrices I & II. Chelsea (New York).

Eising R., 1982: "The Distance Between a System and the Set of Uncontrollable Systems". memo COSOR 82-19, Eindhoven Univ. of Technology.

Eising, R., 1984: "Between Controllable and Uncontrollable". Systems & Control Letters, Vol. 4,n. 5 pp. 263-264, July 1984.

Klema, V.C. and A.J. Laub, 1980: The Singular Value Decomposition: its Computation and Some Applications. IEEE Trans. Automatic Control, Vol. AC-25, no. 2, pp. 164-167.

Kalman, R.E., P. Falb and M.A. Arbib, 1969: Topics in Mathematica System Theory. McGraw-Hill.

Kublanovskaya, V.N., 1961: On Some Algorithms for the Solution of the Complete Eigenvalue Problem. Zh. Vych. Mat., Vol.1, pp. 555-570.

Kublanovskaya, V.N., 1968: On a Method for Solving the Complete Eigenvalue Problem for a Degenerate Matrix. USSR Computational Math. and Math. Phys. Vol. 6, pp. 1-14.

Lawson, C. and R. Hanson, 1974: Solving Linear Least Squares Problems. Prentice-Hall.

Miminis, G., 1981: Numerical Algorithms for Controllability and Eigenvalue Allocation. M. Sc. Thesis, McGill University.

Paige, C.C., 1981: Properties of Numerical Algorithms Related to Computing Controllability. IEEE Trans. Automatic Control, Vol. AC-26, no. 1, pp. 130-138.

Smith, B.T., et al., 1976: Matrix Eigensystems Routines - EISPACK Guide. Lecture Notes in Computer Science 6, Springer-Verlag (Berlin).

Stewart, G.W., 1973: Introduction to Matrix Computations. Academic Press.

Van Dooren P., A.Emani-Naeini and L.Silverman, 1979: Stable Extraction of the Kronecker Structure of Pencils. Proc. 17th IEEE Conference on Decision and Control, pp. 521-524.

Van Dooren, P., 1979: The Computation of Kronecker's Canonical Form of a Singular Pencil. Linear Algebra and Applications, Vol.27, pp. 103-141.

Van Dooren, P., 1981: The Generalized Eigenstructure in Linear Systems Theory. IEEE Trans. Automatic Control, Vol. AC-26, no.1, pp. 111-130.

Wilkinson, J.H., 1965: The Algebraic Eigenvalue Problem. Claredon Press (Oxford).

Wilkinson, J.H. and C.Reinsch, 1971: Linear Algebra - Handbook for Automatic Computation. Vol.2, Springer-Verlag (Berlin).

Chapter 7

SOME RECENT DEVELOPMENTS IN ECONOMETRICS

Michael McALEER and Manfred DEISTLER *

I. INTRODUCTION

Econometrics, in a wide sense, is concerned with the application of statistical or mathematical methods to the analysis of economic phenomena. In this sense, econometrics may be thought of as consisting of the following four fields:

(i) Economic statistics: problems of definition of economic variables (such as in National Income and Product Accounts), problems of data collection, sampling and data construction, and problems of validating the data;

(ii) Econometrics in the narrow sense: econometric methods and econometric model building;

(iii) Economic theories: the use of mathematical formulations;

(iv) Econometric computing: data bank systems for economic data and computer programs, and interactive computing systems for data transformation, estimation and test procedures, graphical displays, calculation of solutions, and computer simulation.

We will be mainly concerned with econometric methods here. Econometrics was born in the 1930's, from the evolving (Keynesian) business cycle theories and the first national accounts, and was especially advanced by the statistical methods developed by the Cowles Commission.

* The authors are grateful to Dr. A. Pagan (Canberra) for valuable comments.

The first econometric models for national economies were due to Tinbergen and Klein, and were built in the forties. In the sixties, for almost every industrialized country, macroeconometric models had been established. These models, and especially forecasts based on these models, behaved fairly well in the periods of steady economic growth, but showed a relatively poor performance in trying to cope with the economic problems arising from the seventies. This poor performance of econometric models had great implications for the standing of econometrics, but the attempt to overcome these difficulties has been one of the main driving forces for the development of current econometrics.

In analyzing these problems, it was found that many of the "a priori" assumptions which had been used in traditional econometric model building, such as those concerning the classification of variables as endogenous and exogenous, the functional form of the relation between the variables, the dynamic specification of the model, and the correlation structure of the errors, could hardly be justified on the basis of real economic a priori knowledge, and that these assumptions had been imposed primarily for statistical convenience. Moreover, the fact that often, by using different a priori assumptions, different conclusions from the same or from similar data sets could be derived, showed that econometrics was far from obtaining objective results from data. The consequence was a critical re-examination of traditional methods and of the assumptions justifying them, and the development of more appropriate methods and research strategies that were more closely related to the actual problems arising in economics.

A criticism that has frequently been raised is that there is a great discrepancy between the process of actually drawing conclusions from economic data, and inference, as described by the decision theoretically-oriented mathematical theory of statistics (see e.g. Leamer (1978)). In many applications the situation is far too complicated to express a statistical decision with one formula. Learning from data may consist of several steps where subjective decisions cannot be excluded at each stage. This was paralleled by the development of exploratory data analysis, as reported in the seminal work of Mosteller and Tukey (1977) in (general) statistics.

Special emphasis has recently been directed in econometrics towards developing methods for checking the model specification from the data, and on data-oriented specification search procedures. In particular, a great number of tests and diagnostic checks have been developed in the last fifteen years to detect misspecification of different kinds (see Pagan and Hall (1983) for a useful discussion of many of these developments). Information criteria have also been developed and used to determine automatically the dynamic specification of various lags of models. A further development concerns the performance of estimators or tests if the data generating process is not described in the model class, and this area of (potentially) misspecified models has recently been investigated by Kent (1982) and White (1982). Two additional areas of current research interest are the robustness of estimators and tests to departures from the assumptions made in using models, as well as the sensitivity of inferences drawn to changes in the assumptions and differences in a priori information.

Until the late sixties, a good part of econometric model-building activity was concerned with macroeconomic modelling and forecasting. Since then, there has been an increasing number of econometric investigations on a far less aggregated level which has led to new models and methods. Moreover, owing to the increased quantity and improved quality of data available, applications of more data-consuming techniques have increased. These "microeconometric" methods are definitely among the most important developments in econometrics today, and we will describe them briefly in this paper.

The question of appropriate macroeconomic modelling and forecasting is still an unresolved issue, after the difficulties encountered in traditional structural model building. Several proposals have been made to overcome these difficulties associated with the traditional approach, and we will describe some of the most important developments in this paper.

The paper is organized as follows. Section 2 is concerned with the specification and quality control of models, and the related issue of specification searches. Macroeconomic modelling and forecasting is discussed in Section 3, and some examples of modern "microeconometric" models are given in Section 4. Needless to say, our account is far from complete and a number of important topics have been omitted. In particular, the (relatively) inexpensive role of computers in econometrics, such as for conducting sophisticated Monte Carlo experiments for comparing different estimators and different test statistics, and for bootstrapping the small sample distributions of estimators and test statistics, is not discussed although they will play a very important role in the development of the discipline in the decades to come.

2. SPECIFICATION AND QUALITY CONTROL OF A MODEL

Differences in formulating a model have been described by McAleer and Pesaran (1986) as follows: differences in theoretical paradigms, differences in the way that auxiliary assumptions within a paradigm are specified, or different strategies that might be adopted in the process of model construction.

By a model specification is meant the set of all assumptions which define the model class, and hence also the parameter space for the inference procedure. Economic theory, e.g., often suggests the variables in a relationship, but not the appropriate functional form or the direct links between the various parts of a system. For these reasons, a data-oriented specification search procedure is warranted, where by specification search is meant the set of procedures followed in moving from an initial model specification to a final model class.

Two matters which arise when there are conflicting views regarding assumptions are the justification of the assumptions from the data and the effect of altering any of them on the properties of inference procedures. The former issue has to do, e.g., with hypothesis testing and diagnostic evaluation, whereas the latter is concerned with robustness of inference procedures to changes in the underlying assumptions.

There are several ways of conducting a specification search, and they may be given as follows:

(i) Data analytic methods: these procedures are based on recognizing patterns in data, as well as in their transformations, and rely heavily on subjective decisions. A well known example of this approach is the method advocated by Box and Jenkins (1970).

(ii) Information criteria, or criteria which provide a trade-off between goodness-of-fit and the number of parameters used to obtain this fit for different model specifications, that is, for different candidates in the specification search.

(iii) Testing procedures: this would appear to be the most common specification search procedure in econometrics. This is, in fact, the third of five stages used in the quality control of a model, as outlined by McAleer, Pagan and Volker (1985). The other stages are given as: checking consistency with economic theory; economic and statistical considerations, such as signs and magnitudes of estimated coefficients, as well as statistical significance; sensitivity analysis; and reconciliation of empirical findings with the results obtained from previous research using alternative non-nested models.

2.1 Model Specification

In what follows, we will concentrate on the role of diagnostic checking within the framework of the multiple linear regression model $y = X\beta + u$, $u \sim N(0, \sigma^2 I)$, which may be written for the t'th observation as

$$y_t = x_t'\beta + u_t \quad , \qquad u_t \sim NID(0, \sigma^2) \tag{2.1}$$

where y_t is the dependent variable, x_t' is the t'th row of the $T \times k$ observation matrix X comprising T observations on k regressors, and u_t is the random error that is assumed to be normally distributed with zero mean, identically and independently distributed for all observations $t = 1,2,\ldots,T$, and uncorrelated with X.

It should be noted that virtually all of the assumptions made in the context of the model given above are, in fact, testable. The properties of the error term, namely, zero mean, serial independence, homoscedasticity and normality, are testable using what are by now standard procedures. Assumptions regarding linearity of the model, a correctly specified set of explanatory variables, and constancy of the coefficients are also all testable. Finally, the informational content of the model given in equation (2.1) may be reconciled with the empirical findings of previous research by recourse to recently developed non-nested testing procedures.

2.2 Tight and Loose Specifications

Before proceeding, it will be necessary to discuss briefly two alternative approaches to specification searches, namely, tight and loose specifications, with corresponding tests of misspecification and specification.

In a tight specification, a very small model set is specified as a first step, and then a series of diagnostic checks is used to indicate ways in which it may be respecified by enlargening the model set. Diagnostic checks are primarily tests of misspecification since only the null hypothesis needs to be specified in advance of performing the test. Consider the following examples: testing for serial independence of errors against either AR(p) or MA(p) alternatives can result in the same test statistic (Godfrey (1978)); testing for homoscedasticity of the errors against either multiplicative or additive heteroscedasticity as alternatives can also result in the same test statistic (see Bera and McKenzie (1986)).

Bearing the caveats given above in mind, rejection of the null hypothesis using diagnostic checks *may* suggest where to look to improve the model specification (see Table 1).

The scheme given in Table 1 should be used regardless of whether or not a tight specification is used. However, the more tight the specification, the more likely it is that instances will be found where the diagnostic checks lead to rejection of the null hypothesis.

If there are many observations available, it may be useful to commence with a very large model set and to test restrictions on that loose specification. In situations such as this, in which both the null and alternative hypotheses are specified, a test of specification is being considered. An important approach to consider in testing restrictions on a loosely specified model class using time series data is that of uniquely ordered hypotheses (see Anderson (1971)). In this approach, if any hypothesis is rejected, any succeeding hypothesis is also rejected and need not be tested.
It is advisable to start with a (fairly) general hypothesis (i.e. the maintained hypothesis) and to test hypotheses in increasing order of restrictiveness until a rejection occurs, or the most restrictive hypothesis is accepted (i.e. is not rejected). The accepted hypothesis is the one prior to rejection. An interesting and useful application of uniquely ordered hypotheses in econometrics is that of testing for common factors (see Sargan (1980) for theoretical considerations, and Hendry and Mizon (1978) for an illustration).

TABLE 1

Using Diagnostic Checks to Test for Possible Model Misspecification

Diagnostic check	Possible sources of error
Serial correlation	Correlated errors Omitted variables Incorrect functional form Incorrect transformation of variables
Heteroscedasticity	Non-constant variances Incorrect functional form Incorrectly transformed dependent variable
Exogeneity	Measurement errors Omitted links with larger system
Functional form	Omitted variables Incorrect transformations on variables Incorrect functional form
Parameter constancy	Structural change Varying coefficients Weak forecasting ability
Non-nested alternatives	Incorrect model Alternative explanations possible

2.3 Principles for Testing

Returning now to the model given in equation (2.1), let us denote the ordinary least squares (OLS) estimators of β and σ^2 as $\hat{\beta} = (X'X)^{-1}X'y$ and $\hat{\sigma}^2 = (y-X\hat{\beta})'(y-X\hat{\beta})/(T-k)$. The OLS residuals are given as $\hat{u} = y-X\hat{\beta}$, with t'th comment given by $\hat{u}_t = y_t - x_t'\hat{\beta}$.

Tests may be constructed using the following Principles: The Likelihood Ratio, Wald and Lagrange Multiplier (or Score) Principles (see Engle (1983) for a discussion); the Cox (1961, 1962) Principle for non-nested (or separate) families of hypotheses (see McAleer and Pesaran (1986)); and the test procedures based on the work of Durbin (1954) and Hausman (1978) (see Ruud (1984) for further details). The first three Principles lead to tests which are asymptotically equivalent under the null hypothesis as well as under local alternatives, and the likelihood ratio test is the only one of the three which requires estimation under both the null and the alternative hypothesis. The Lagrange multiplier (LM) test is extremely straightforward to use computationally, as it can frequently be computed as TR^2, that is, the sample size times the coefficient of multiple determination from an auxiliary linear regression. Several examples will be given below to illustrate the simplicity of the LM procedure.

The Cox Principle is a general approach that may be used for testing non-nested hypotheses, a special case of which is that of nested hypotheses. It essentially involves centring any given test statistic under the null hypothesis, and then deriving its asymptotic null distribution. This procedure can be applied to the likelihood ratio statistic itself, or to residual sums of squares from different regression models, or to differences in some or all of the parameter estimates of alternative models. The Hausman test procedure (Hausman (1978)) can be considered to be an application of the Cox Principle. This procedure is based on the difference between two estimators, one of which is efficient under the null, but not even consistent under the alternative hypothesis, whereas the other is consistent regardless of whether the null or alternative hypothesis is true.

2.4 Diagnostic Testing

Throughout this section it will be presumed that a regression package is available for providing OLS estimates and for storing the predictions and residuals from OLS estimation. Unless stated otherwise, all test statistics discussed below may be obtained from OLS regressions based on simple auxiliary equations. Emphasis is placed on non-structural models, and the interested reader is referred to the review by Pagan and Hall (1983) for a detailed discussion of extensions to structural models. Since the following discussion is necessarily limited in scope, the broader treatment provided by Pagan (1984) is also highly recommended.

2.4.1 Serial correlation

Serial correlation of the error term u_t can lead to inefficient estimators and predictions, and to inconsistent estimates of β if a lagged dependent variable is in the set of regressors. Perhaps the most useful tests for serial independence against AR(p) or MA(p) alternatives have been developed by Durbin (1970), see also Breusch (1978) and Godfrey (1978). If u_t follows an AR(p) process, then $u_t = \rho_1 u_{t-1} + \rho_2 u_{t-2} + \dots + \rho_p u_{t-p} + \varepsilon_t$, where ε_t is white noise, and the regression model is given by $y_t = x_t'\beta + \rho_1 u_{t-1} + \rho_2 u_{t-2} + \dots + \rho_p u_{t-p} + \varepsilon_t$. The LM test of the null hypothesis H_o: $\rho_1 = \rho_2 = \dots = \rho_p = 0$ is obtained by replacing u_{t-j} with $\hat{u}_{t-j}$ $(j = 1,2,\dots,p)$, the lagged values of the OLS residuals. The LM test statistic is calculated as TR^2 from the auxiliary regression

$y_t = x_t'\beta + \rho_1\hat{u}_{t-1} + \rho_2\hat{u}_{t-2} + \ldots + \rho_p\hat{u}_{t-p} + \varepsilon_t^*$, with TR^2 distributed asymptotically as $\chi^2(p)$ under the null hypothesis. The LM test for serial independence against an MA(p) process is given by the same auxiliary regression. Setting $p = 1$ gives an asymptotically equivalent test to Durbin's (1970) h-statistic.

2.4.2 Heteroscedasticity

When the variance of the error term is not constant but varies with each observation, we might think of it as having the relation $\sigma_t^2 = \sigma^2 + z_t'\gamma$, where z_t is given. The LM test statistic is calculated by regressing the squared OLS residuals, $\hat{u}_t^2$, on a constant and a vector of variables. For example, if z_t is given as the scalar $E(y_t)$ or $\ell n\, E(y_t)$, the LM statistic is obtained as TR^2 from the auxiliary regression $\hat{u}_t^2 = \alpha + \gamma\hat{y}_t + \upsilon_t$ or $\hat{u}_t^2 = \alpha + \gamma \ell n\, \hat{y}_t + \upsilon_t$, where υ_t is the equation error. In cases where the explicit form of heteroscedasticity might be suspected, this could be incorporated into the construction of the vector z_t'.

2.4.3 Exogeneity

In the model $y = X\beta + u$, lack of exogeneity of X, either through measurement errors or because some elements of X are determined within a larger system, may lead to inconsistent estimators of β. The Hausman test for exogeneity can be applied straightforwardly, as discussed in Section 2.3. A computationally convenient method for calculating the test is to use a set of instrumental variables W and to obtain the predictions $\hat{X} = W(W'W)^{-1}W'X$ from regressing the columns of X on those of W. The auxiliary regression $y = X\beta + \hat{X}\psi + u$ is estimated to test the null hypothesis of exogeneity $H_o: \psi = 0$, where the F statistic is asymptotically valid for testing H_o (see also Durbin (1954)).

2.4.4 Functional form

The most straightforward diagnostic check for omitted variables and/or incorrect functional form is undoubtedly Ramsey's (1969) regression specification error test (RESET). This involves adding powers of the predictions from the null model to the system and testing for the presence of the additional factors. The augmented regression is, for example, $y_t = x_t'\beta + \gamma_1\hat{y}_t^2 + \gamma_2\hat{y}_t^3 + u_t$ and the F test of $H_o: \gamma_1 = \gamma_2 = 0$ is distributed exactly as $F(2,T-k-2)$ under H_o if the regressors x_t are exogenous.

Several additional tests for incorrect functional form are available, and some of these methods have been based on the data transformations suggested by Box and Cox (1964). In particular, Andrews (1971) has derived a linearized version of the Box-Cox model which does not require estimation of the Box-Cox model itself, but only the specialization of it that is being tested. Godfrey and Wickens (1981) have derived LM tests of both linear and log-linear specifications against the more general Box-Cox model that are calculated as TR^2 from auxiliary regressions. For a survey of alternative tests of linear and log-linear models, as well as a discussion of their small sample properties, see McAleer (1985, Section 6).

2.4.5 Parameter constancy

Perhaps the most well known test for parameter constancy in econometrics is the Chow test, in which the null hypothesis of constancy of parameters in the linear regression model is tested against the alternative of a change in coefficients at a known point in time. The cumulative sum and cumulative sum of squares tests of Brown, Durbin and Evans (1975), which are based on recursive residuals, is available when a broader class of parameter non-constancies is considered as an alternative. The constancy of parameters may also be checked by testing for predictive ability based upon post-sample observations (see Salkever (1976) for a very simple test for parameter constancy). A useful review of this literature is given in Pesaran, Smith and Yeo (1985).

2.4.6 Non-nested alternatives

The result of applying diagnostic checks to various model specifications may lead to two or more models that are not rejected. When one model cannot be obtained from another by the imposition of restrictions, the models are said to be non-nested. The most well known test, namely the Cox test, was introduced into the econometric literature by Pesaran (1974). Let the null hypothesis be given by H_o: $y = X\beta + u$, and let the alternative be H_1: $y = Z\gamma + v$. The simplest test of H_o against H_1 is the test of H_o: $\gamma = 0$ in the augmented regression $y = X\beta + Z\gamma + u$, and this has been justified in the literature as testing "parameters of interest", as an encompassing test, and as a test based on Roy's Union-intersection Principle (see McAleer and Pesaran (1986) for details). Two other Cox-type tests are given in Davidson and MacKinnon (1981) and Fisher and McAleer (1981). All of these tests are compared with regard to small sample properties in McAleer (1985, Section 7).

3. MACROECONOMIC MODELLING AND FORECASTING

Macroeconomic modelling is concerned with modelling the dynamics of, and the interaction between, highly aggregated macroeconomic variables such as national income, consumption, investment, unemployment and prices; of special interest is the study of the business cycle, forecasting and policy simulation.

The traditional approach to macroeconomic modelling is structural model building, when large-scale models comprised e.g. of a hundred or more equations are estimated from the data and used for economic analysis forecasting and policy simulation. In view of the relatively large number of equations and the relatively small data sets involved, a great number of restrictions on the parameters have to be imposed to obtain reasonably low-dimensioned parameter spaces and reasonable parameter estimates. These restrictions are frequently in the form of zero restrictions, indicating that a certain variable does not influence some other variable in a certain equation of the system. Under the assumption of an a priori given specification, the theory of identifiability and maximum likelihood estimation of linear simultaneous equation systems (with uncorrelated errors) was developed by Koopmans, Rubin and Leipnik (1950); two-stage and three-stage least squares methods were subsequently developed in order to simplify calculations. However, in practical applications the most common estimation method was, and still is, ordinary least squares, despite its lack of consistency in the simultaneous equation framework.

In the period of steady economic growth, the traditional large scale models showed satisfactory forecasting behaviour. But with the increasing fluctuations in many economic variables after the oil shock of the seventies, many of these models showed rather poor forecasting performances. At this time the first macroeconomic forecasts were made with Box-Jenkins models and these simple univariate models often out performed large econometric models, at least for short-term forecasts.

As has been said previously, these facts led to a widespread critique of the traditional model-building approach. One of the main reasons for the weakness of many forecasts was found in the poor specification of the respective models, where too much economic a priori information was presumed to be available. As a consequence, data-oriented specification search procedures (as described in Section 2) and new models have been proposed.

At present, different modelling philosophies, and hence different models, have been proposed, even for identical or similar data sets. Therefore, macroeconometric modelling is still far from lacking in controversy. There are still advocates of traditional structural model-building, especially if the main aim is an analysis of the interaction between variables and policy simulation, rather than (unconditional) forecasting; on the other hand, several different proposals for new modelling approaches have been made, and we will describe some of these below.

A method for obtaining the dynamic specification of a structural model is described in Zellner and Palm (1974) as follows. Let

$$\sum_{i=0}^{p} A_i y_{t-i} = \sum_{i=0}^{q} B_i z_{t-i} + u_t \tag{3.1}$$

be the structural model, where $A_i \in \mathbb{R}^{s\times s}$, $B_i \in \mathbb{R}^{s\times m}$ are the parameter matrices, and y_t, z_t and u_t are the endogenous, exogenous and white noise error variables, respectively. Equation (3.1) may be transformed by a left multiplication by the adjoint of $\Sigma A_i B^i$ (where B is the backward-shift operator) to yield a system which is decoupled in the sense that only the i'th endogeneous variable (including its lagged values) appears in its i'th equation. In this form, the equations can be treated as s single equations andthe Box-Jenkins method can be applied to obtain the dynamic specification for each of these single equations. The dynamic specification of the original model is then obtained from these specifications. Of course, one problem associated with this procedure is that the zero restrictions of the original model are not taken into account in the transformed single equations.

The classification of the observed variables as endogeneous and exogenous often cannot be justified on a priori grounds; this is especially true if conflicting economic theories imply different classifications for the variables. This ambiguity has led to discussions concerning the concept of exogeneity and to tests for exogeneity. One concept of exogeneity is related to causality in the sense introduced by Granger (1969). In this analysis,variables are called exogenous if there is a unidirectional causal influence from the exogenous to the endogenous variables. Tests for causality have been proposed, e.g. by Sims (1972), and Pierce and Haugh (1977). The second concept of exogeneity was given in Engle, Hendry and Richard (1983). The defining property of exogeneity here is that conditioning the other observed variables on the exogenous variables gives no loss of information about the parameters of interest.

Another approach to overcome the classification problem discussed above is to provide a symmetric treatment to all observed variables, by describing them jointly as a vector autoregressive (VAR) process. This has been proposed e.g. in Sims (1980). Once the VAR system has been estimated, questions such as the classification of variables into exogenous and endogenous, or whether there are zero restrictions among the parameters, may be answered on an empirical basis (see Sargent and Sims (1977), and Hsiao (1982)). Although these vector autoregressions usually contain significantly fewer equations (about ten) than the usual structural models, both the dimension of the parameter space as well as the dynamic specification of the model remain problematic.

In order to reduce the dimensions of the parameter space, Sargent and Sims (1977) have proposed a dynamic principal component analysis where the dynamics in the observed variables are introduced by factors of lower dimension; this is very closely related to the idea that the business cycle in most macroeconomic variables can be explained by a few dynamic factors (Bowden (1972)). However, in determining the number of dynamic factors empirically, there seems to be no sharp delineation of principal components (Sims,private communication).

A Bayesian procedure for analyzing VAR models has been proposed in Doan Litterman and Sims (1984), which seems to have surprisingly good forecasting properties. In this approach the prior means of all coefficients corresponding to lags greater than one are set equal to zero and the estimation problem is reduced to the estimation of relatively few "hyperparameters", such as the tightness of the prior means.

Another method of reducing the dimension of VAR systems is the use of the AIC or BIC criteria to determine both the maximum lags in the VAR model and zero restrictions on the coefficients corresponding to smaller lags. For multivariate subset autoregressive modelling, see e.g. Penm and Terrell (1984). This method also performs well in forecasting.

A different approach is to concentrate on the modelling of one equation at a time (see Davidson et al.(1978) and Hendry (1986)) which implicitly assumes that the effects of simultaneity are negligible. This approach stresses that economic a priori knowledge is primarily concerned with the long-run equilibrium solutions of the system, whereas in many cases economic theory has very little to say about short-run behaviour. For this reason, equilibrium solutions of dynamic models should be consistent with economic theory.

4. MICROECONOMETRICS

During the last decade there has been a substantial development of econometric techniques to answer questions posed in empirical microeconomics. As an important example, consider the case where the variable being explained by the model either takes on only discrete values, or is limited in its range. Sample survey data frequently requires such models to be used: for example, a binary-choice model might be used to explain the decision to buy a car or not, and a multiple-choice model might be used to explain whether a commuter travels by bus, train or car. The explanatory variables in each of these examples would be the personal and economic characteristics of various individuals.

The conditional probabilities of the outcomes of the discrete variable are related to various explanatory variables in the model. Owing to the nature of probabilities, the functional form of these relations must be restricted; in particular, linear relations are excluded. In binary models, where there is only one conditional probability to be explained, the most important class of models is of the form $P(E|x_t'\beta) = F(x_t'\beta)$, where E denotes outcome of one event and F is a cumulative distribution function. The most frequently used functions are the normal and the logistic, respectively, leading to probit and logit models. The most well known model where the dependent variable is limited in its range is the Tobit model (Tobin (1958)). For example, many observations in a sample may take on the value zero (if, say, it is decided not to buy a car) while other individuals may spend varying amounts on cars. In this sense, the dependent variable is part qualitative and part quantitative. The standard method of estimation of these models is by maximum likelihood. For further details, see Amemiya (1981), Maddala (1983) and McFadden (1976).

REFERENCES

Amemiya, T. (1981): Qualitative Response Models : A Survey. Journal of Economic Literature 19, 1483-1536.

Anderson, T.W. (1971): The Statistical Analysis of Time Series. Wiley, New York.

Andrews, D.F. (1971): A Note on the Selection of Data Transformations. Biometrika 58, 249-254.

Bera, A.K. and C.R. McKenzie (1986): Alternative Forms and Properties of the Score Test. Forthcoming in Journal of Applied Statistics.

Bowden, R.J. (1972): More Stochastic Properties of the Klein-Goldberger Model. Econometrica 40, 87-98.

Box, G.E.P. and D.R. Cox (1964): An Analysis of Transformations. Journal of the Royal Statistical Society B 26, 211-252.

Box, G.E.P. and G.M. Jenkins (1970): Time Series Analysis, Forecasting and Control. Holden Day, San Francisco.

Breusch, T.S. (1978): Testing for Autocorrelation in Dynamic Linear Models. Australian Economic Papers 17, 334-355.

Brown, R.L., J. Durbin and J.M. Evans (1975): Techniques for Testing the Constancy of Regression Relationships Over Time. Journal of the Royal Statistical Society B 37, 149-192.

Chow, G.C. (1960): Tests of Equality Between Sets of Coefficients in Two Linear Regressions. Econometrica 28, 591-605.

Cox, D.R. (1961): Tests of Separate Families of Hypotheses. In: Proceedings of the Fourth Berkeley Symposium on Mathematical Statistics and Probability 1. University of California Press, Berkeley.

Cox, D.R. (1962): Further Results on Tests of Separate Families of Hypotheses. Journal of the Royal Statistical Society B 24, 406-424.

Davidson, J.E.H., D.F. Hendry, F. Srba and J.S. Yeo (1978): Econometric Modelling of the Aggregate Time-Series Relationship Between Consumers' Expenditure and Income in the United Kingdom. Economic Journal 88, 661-692.

Davidson, R. and J. MacKinnon (1981): Several Tests for Model Specification in the Presence of Alternative Hypotheses. Econometrica 49, 781-793.

Doan, T., R. Litterman and C. Sims (1984): Forecasting and Conditional Projection Using Realistic Prior Distributions. Econometric Reviews 3, 1-100.

Durbin, J. (1954): Errors in Variables. Review of the International Statistical Institute 22, 23-32.

Durbin, J. (1970): Testing for Serial Correlation in Least Squares Regression When Some of the Regressors are Lagged Dependent Variables. Econometrica 38, 410-421.

Engle, R.F. (1983): Wald, Likelihood Ratio and Lagrange Multiplier Tests in Econometrics. In: Z. Griliches and M. Intriligator (Eds.) Handbook of Econometrics. North-Holland, Amsterdam.

Engle, R.F., D.F. Hendry and J-F. Richard (1983): Exogeneity. Econometrica 51, 277-304.

Fisher, G. and M. McAleer (1981): Alternative Procedures and Associated Tests of Significance for Non-Nested Hypotheses. Journal of Econometrics 16, 103-119.

Godfrey, L.G. (1978): Testing for Higher Order Serial Correlation in Regression Equations When the Regressors Include Lagged Dependent Variables. Econometrica 46, 1303-1310.

Godfrey, L.G. and M.R. Wickens (1981): Testing Linear and Log-Linear Regressions for Functional Form. Review of Economic Studies 48, 487-496.

Granger, C.W.J. (1969): Investigating Causal Relationships by Econometric Models and Cross-Spectral Methods. Econometrica 37, 424-438.

Hausman, J.A. (1978): Specification Tests in Econometrics. Econometrica 46, 1251-1271.

Hendry, D.F. (1986): Empirical Modelling in Dynamic Economics. Forthcoming in Applied Mathematics and Computation.

Hendry, D.F. and G.E. Mizon (1978): Serial Correlation as a Convenient Simplification, Not a Nuisance : A Comment on a Study of the Demand for Money by the Bank of England. Economic Journal 88, 549-563.

Hsiao, C. (1982): Autoregressive Modelling and Causal Ordering of Economic Variables. Journal of Economic Dynamics and Control 4, 243-259.

Kent, J.T. (1982): Robust Properties of Likelihood Ratio Tests. Biometrika 69, 19-27.

Koopmans, T.C., H. Rubin and R.B. Leipnik (1950): Measuring the Equation Systems of Dynamic Economics. In: T.C. Koopmans (Ed.) Statistical Inference in Dynamic Economic Models. Wiley, New York.

Leamer, E.E. (1978): Specification Searches : Ad Hoc Inference with Non-experimental Data. Wiley, New York.

Maddala, G.S. (1983): Limited Dependent and Qualitative Variables in Econometrics. Cambridge University Press.

McAleer, M. (1985): Specification Tests for Separate Models: A Survey. In: M.L. King and D.E.A. Giles (Eds.) Specification Analysis in the Linear Model. Routledge and Kegan Paul, London.

McAleer, M., A.R. Pagan and P.A. Volker (1985): What Will Take the Con Out of Econometrics? American Economic Review 75, 293-307.

McAleer, M. and M.H. Pesaran (1986): Statistical Inference in Non-nested Econometric Models. Forthcoming in Applied Mathematics and Computation.

McFadden, D. (1976): Quantal Choice Analysis : A Survey. Annals of Economic and Social Measurement 5, 363-390.

Mosteller, F. and J.W. Tukey (1977): Data Analysis and Regression. Addison-Wesley, New York.

Pagan, A.R. (1984): Model Evaluation by Variable Addition. In: D.F. Hendry and K.F. Wallis (Eds.) Econometrics and Quantitative Economics. Blackwell, Oxford.

Pagan, A.R. and A.D. Hall (1983): Diagnostic Tests as Residual Analysis. Econometric Reviews 2, 159-218.

Penm, J.H.W. and R.D. Terrell (1984): Multivariate Subset Autoregressive Modelling with Zero Constraints for Detecting "Overall Causality". Journal of Econometrics 24, 311-330.

Pesaran, M.H. (1974): On the General Problem of Model Selection. Review of Economic Studies 41, 153-171.

Pesaran, M.H., R.P. Smith and J.S. Yeo (1985): Testing for Structural Stability and Predictive Failure: A Review. The Manchester School 53, 280-295.

Pierce, D.A. and L.D. Haugh (1977): Causality in Temporal Systems. Journal of Econometrics 5, 265-293.

Ramsey, J.B. (1969): Tests for Specification Errors in Classical Linear Least Squares Regression Analysis. Journal of the Royal Statistical Society B 31, 350-371.

Ruud, P.A. (1984): Tests of Specification in Econometrics. Econometric Reviews 3, 211-242.

Salkever, D.S. (1976): The Use of Dummy Variables to Compute Predictions, Prediction Errors and Confidence Intervals. Journal of Econometrics 4, 393-397.

Sargan, J.D. (1980): Some Tests of Dynamic Specification for a Single Equation. Econometrica 48, 879-897.

Sargent, T.J. and C.A. Sims (1977): Business Cycle Modelling Without Pretending to Have Too Much A Priori Economic Theory. In: C.A. Sims (Ed.) New Methods of Business Cycle Research. Federal Reserve Bank of Minneapolis.

Sims, C.A. (1972): Money, Income and Causality. American Economic Review 62, 540-552.

Sims, C.A. (1980): Macroeconomics and Reality. Econometrica 48, 1-48.

Tobin, J. (1958): The Estimation of Relationships for Limited Dependent Variables. Econometrica 26, 24-36.

White, H. (1982): Maximum Likelihood Estimation of Misspecified Models. Econometrica 50, 1-25.

Zellner, A. and F. Palm (1974): Time Series Analysis and Simultaneous Equation Econometric Models. Journal of Econometrics 2, 17-54.

Lecture Notes in Control and Information Sciences

Edited by M. Thoma

Vol. 43: Stochastic Differential Systems
Proceedings of the 2nd Bad Honnef Conference
of the SFB 72 of the DFG at the University of Bonn
June 28 – July 2, 1982
Edited by M. Kohlmann and N. Christopeit
XII, 377 pages. 1982.

Vol. 44: Analysis and Optimization of Systems
Proceedings of the Fifth International
Conference on Analysis and Optimization of Systems
Versailles. December 14–17, 1982
Edited by A. Bensoussan and J. L. Lions
XV, 987 pages, 1982

Vol. 45: M. Arató
Linear Stochastic Systems
with Constant Coefficients
A Statistical Approach
IX, 309 pages. 1982

Vol. 46: Time-Scale Modeling of Dynamic Networks
with Applications to Power Systems
Edited by J. H. Chow
X, 218 pages. 1982

Vol. 47: P. A. Ioannou, P. V. Kokotovic
Adaptive Systems with Reduced Models
V, 162 pages. 1983

Vol. 48: Yaakov Yavin
Feedback Strategies for Partially
Observable Stochastic Systems
VI, 233 pages, 1983

Vol. 49: Theory and Application of Random Fields
Proceedings of the IFIP-WG 7/1
Working Conference
held under the joint auspices of the
Indian Statistical Institute
Bangalore, India, January 1982
Edited by G. Kallianpur
VI. 290 pages. 1983

Vol. 50: M. Papageorgiou
Applications of Automatic Control Concepts
to Traffic Flow Modeling and Control
IX, 186 pages. 1983

Vol. 51: Z. Nahorski, H.F. Ravn, R.V.V. Vidal
Optimization of Discrete Time Systems
The Upper Boundary Approach
V, 137 pages 1983

Vol. 52: A. L. Dontchev
Perturbations, Approximations and Sensitivity Analysis
of Optimal Control Systems
IV, 158 pages. 1983

Vol. 53: Liu Chen Hui
General Decoupling Theory of Multivariable
Process Control Systems
XI, 474 pages. 1983

Vol. 54: Control Theory for Distributed
Parameter Systems and Applications
Edited by F. Kappel, K. Kunisch,
W. Schappacher
VII, 245 pages. 1983.

Vol. 55: Ganti Prasada Rao
Piecewise Constant Orthogonal Functions
and Their Application to Systems and Control
VII, 254 pages. 1983.

Vol. 56: Dines Chandra Saha, Ganti Prasada R
Identification of Continuous
Dynamical Systems
The Poisson Moment Functional
(PMF) Approach
IX, 158 pages. 1983.

Vol. 57: T. Söderström, P. G. Stoica
Instrumental Variable Methods
for System Identification
VII, 243 pages. 1983.

Vol. 58: Mathematical Theory of
Networks and Systems
Proceedings of the MTNS-83 International
Symposium
Beer Sheva, Israel, June 20–24, 1983
Edited by P. A. Fuhrmann
X, 906 pages. 1984

Vol. 59: System Modelling and Optimization
Proceedings of the 11th IFIP Conference
Copenhagen, Denmark, July 25-29, 1983
Edited by P. Thoft-Christensen
IX, 892 pages. 1984

Vol. 60: Modelling and Performance
Evaluation Methodology
Proceedings of the International Seminar
Paris, France, January 24–26, 1983
Edited by F. Bacelli and G. Fayolle
VII, 655 pages. 1984

Vol. 61: Filtering and Control of Random
Processes
Proceedings of the E.N.S.T.-C.N.E.T. Colloquiur
Paris, France, February 23–24, 1983
Edited by H. Korezlioglu, G. Mazziotto, and
J. Szpirglas
V, 325 pages. 1984

Lecture Notes in Control and Information Sciences

Edited by M. Thoma and A. Wyner

Vol. 62: Analysis and Optimization of Systems
Proceedings of the Sixth International Conference on Analysis and Optimization of Systems
Nice, June 19–22, 1984
Edited by A. Bensoussan, J. L. Lions
XIX, 591 pages. 1984.

Vol. 63: Analysis and Optimization of Systems
Proceedings of the Sixth International Conference on Analysis and Optimization of Systems
Nice, June 19–22, 1984
Edited by A. Bensoussan, J. L. Lions
XIX, 700 pages. 1984.

Vol. 64: Arunabha Bagchi
Stackelberg Differential Games in Economic Models
VIII, 203 pages, 1984

Vol. 65: Yaakov Yavin
Numerical Studies in Nonlinear Filtering
VIII, 273 pages, 1985.

Vol. 66: Systems and Optimization
Proceedings of the Twente Workshop
Enschede, The Netherlands, April 16–18, 1984
Edited by A. Bagchi, H. Th. Jongen
X, 206 pages, 1985.

Vol. 67: Real Time Control of Large Scale Systems
Proceedings of the First European Workshop
University of Patras, Greece, Juli 9–12, 1984
Edited by G. Schmidt, M. Singh, A. Titli, S. Tzafestas
XI, 650 pages, 1985.

Vol. 68: T. Kaczorek
Two-Dimensional Linear Systems
IX, 397 pages, 1985.

Vol. 69: Stochastic Differential Systems – Filtering and Control
Proceedings of the IFIP-WG 7/1 Working Conference
Marseille-Luminy, France, March 12-17, 1984
Edited by M. Metivier, E. Pardoux
X, 310 pages, 1985.

Vol. 70: Uncertainty and Control
Proceedings of a DFVLR International Colloquium
Bonn, Germany, March, 1985
Edited by J. Ackermann
IV, 236 pages, 1985.

Vol. 71: N. Baba
New Topics in Learning Automata Theory and Applications
VII, 231 pages, 1985.

Vol. 72: A. Isidori
Nonlinear Control Systems: An Introduction
VI, 297 pages, 1985.

Vol. 73: J. Zarzycki
Nonlinear Prediction Ladder-Filters for Higher-Order Stochastic Sequences
V, 132 pages, 1985.

Vol. 74: K. Ichikawa
Control System Design based on Exact Model Matching Techniques
VII, 129 pages, 1985.

Vol. 75: Distributed Parameter Systems
Proceedings of the 2nd International Conference, Vorau, Austria 1984
Edited by F. Kappel, K. Kunisch, W. Schappacher
VIII, 460 pages, 1985.

Vol. 76: Stochastic Programming
Edited by F. Archetti, G. Di Pillo, M. Lucertini
V, 285 pages, 1986.

Vol. 77: Detection of Abrupt Changes in Signals and Dynamical Systems
Edited by M. Basseville, A. Benveniste
X, 373 pages, 1986.

Vol. 78: Stochastic Differential Systems
Proceedings of the 3rd Bad Honnef Conference, June 3–7, 1985
Edited by N. Christopeit, K. Helmes, M. Kohlmann
V, 372 pages, 1986.

Vol. 79: Signal Processing for Control
Edited by K. Godfrey, P. Jones
XVIII, 413 pages, 1986.

Vol. 80: Artificial Intelligence and Man-Machine Systems
Edited by H. Winter
IV, 211 pages, 1986.

Lecture Notes in Control and Information Sciences

Edited by M. Thoma and A. Wyner

Vol. 81: Stochastic Optimization
Proceedings of the International Conference,
Kiew, 1984
Edited by I. Arkin, A. Shiraev, R. Wets
X, 754 pages, 1986.

Vol. 82: Analysis and Algorithms
of Optimization Problems
Edited by K. Malanowski, K. Mizukami
VIII, 240 pages, 1986.

Vol. 83: Analysis and Optimization
of Systems
Proceedings of the Seventh International
Conference of Analysis and Optimization
of Systems
Antiba, June 26-27, 1986
Edited by A. Benoussan, J. L. Lions
XVI, 901 pages, 1986.

Vol. 84: System Modelling
and Optimization
Proceedings of the 12th IFIP Conference
Budapest, Hungary, September 2–6, 1985
Edited by A. Prékopa, J. Szelezsán, B. Strazicky
XII, 1046 pages, 1986.

Vol. 85: Stochastic Processes
in Underwater Acoustics
Edited by Charles R. Baker
V, 205 pages, 1986.

Vol. 86: Time Series and
Linear Systems
Edited by Sergio Bittanti
XVII, 243 pages, 1986.